T-Labs Series in Telecommunication Services

For further volumes:
http://www.springer.com/series/10013

Marie-Neige Garcia

Parametric Packet-Based Audiovisual Quality Model for IPTV Services

Springer

Marie-Neige Garcia
Telekom Innovation Laboratories,
Assessment of IP-based
Application (AIPA)
TU Berlin
Berlin
Germany

ISSN 2192-2810 ISSN 2192-2829 (electronic)
ISBN 978-3-319-04854-3 ISBN 978-3-319-04855-0 (eBook)
DOI 10.1007/978-3-319-04855-0
Springer Cham Heidelberg New York Dordrecht London

Library of Congress Control Number: 2014940154

Printed on acid-free paper

Springer is part of Springer Science+Business Media (www.springer.com)

Preface

Today, telecommunication services go beyond telephony and include video services such as Internet Protocol Television. These video services are prone to degradations including blurring, blockiness, or unexpected freezing during video playback. These degradations may be perceived by the user and therefore may spoil her/his experience of the service. As a consequence, the audiovisual quality as perceived by the user needs to be measured and, if needed, improved.

This book presents a parametric comprehensive model to measure and predict the audiovisual quality of Internet Protocol Television services as it is likely to be perceived by the user. The comprehensive model is divided into three sub-models referred to as the audio model, the video model, and the audiovisual model. The audio and video models take as input a parametric description of the audiovisual processing path, and deliver distinct estimates for both the audio and video quality. These distinct estimates are eventually used as input data for the audiovisual model. This model provides an overall estimate of the perceived audiovisual quality in total. The parametric description can be used as diagnostic information.

The quality estimates and diagnostic information can be practically applied to enhance network deployment and operations. Two applications come to mind in particular: Network planning and network service quality monitoring. The audio model can be used indifferently for both applications. However, two variants of the video model have been developed in order to address particular needs of the applications mentioned above.

This book is written for the students, researchers, and engineers involved in the assessment of the Quality of Experience (QoE) of telecommunication services. It may in particular be useful for persons working on network deployment and operations, and for those developing instrumental methods to assess the perceived quality of telecommunication services.

This book was created within the scope of my Ph.D. dissertation, which was conducted under the supervision of Prof. Alexander Raake at the Assessment of IP-based Application (AIPA) chair, Technische Universität Berlin. The work was supported by the T-V-Model project coordinated by Dr. Bernhard Feiten at the Telekom Innovation Laboratories (T-Labs). The scope of this work was also influenced by the ITU-T P.NAMS ("Parametric non-intrusive assessment of audiovisual media streaming quality") competition. The model applied to quality monitoring was indeed submitted during this competition. It resulted in the ITU-T P.1201.2 standard.

Acknowledgments

I owe my deepest gratitude to my colleague, supervisor, and friend Prof. Alexander Raake. He was a guide in my work and scientific career, and pushed any undertaken task to higher scientific levels. He never failed to support me and to share his numerous ideas and his motivation.

I am also indebted to the whole T-V-Model team: Bernhard Feiten, Alexander Raake, Peter List, Savvas Argyropoulos, Ulf Wüstenhagen, and Jens Kroll for their expertise and friendly support. The model presented in this work would not have won the ITU-T P.NAMS competition without them.

In the context of this competition, I would like to thank my colleagues of the ITU-T Study Group 12, in particular Jörgen Gustafsson, David Lindegren, Martin Pettersson, Simon Broom, Kazuhisa Yamagishi, Ning Liao, Simon Gao, Martin Kastner, Akira Takahashi, Jens Berger, and all further ITU-T Study Group 12 colleagues for their involvement in the workgroups Questions Q14/12, Q13/12 and in the Work Party WP2/12.

It was also a great pleasure to participate in Video Quality Expert Group (VQEG) meetings, and I thank the VQEG community for sharing their expertise and for being open-minded to newcomers. I am in particular grateful to Prof. Patrick Le Callet for reviewing my Ph.D. thesis.

I am also thankful to Stephen Wilson, Markus P. Möller, and Rahul Swaminathan for their helpful advice on my Ph.D. dissertation writing and defense.

I would also like to thank my colleagues at the AIPA, Quality and Usability, and Design Research Labs for a unique friendly and scientific atmosphere. It was always a pleasure to come to work.

In particular, I would like to thank:

- Prof. Sebastian Möller, head of the Quality and Usability Lab, for his support, expertise, and useful advice,
- Nicolas Côté, Marcel Wältermann, Dominik Strohmeier, Benjamin Belmudez, Sebastian Egger, Jan-Niklas Antons, and Pierre Lebreton for fruitful discussions on subjective quality assessment and modeling,
- Robert Schleicher, Ina Wechsung, Katrin Schoenenberg, and Tim Polzehl for their expertise and help in statistical analysis and modeling,
- Prof. Sascha Spors and Prof. Marcus Barkowsky for their expertise and useful advice during the set up of standardized usability/quality test rooms,

- Irene Hube-Achter, Christine Kluge, and Ulrike Kehrberg for administrative (and more!) support,
- Katharina Bredies, Benjamin Weiss, Klaus-Peter Engelbrecht, Rahul Swaminathan, Matthias Geier, Christine Kühnel, Matthias Rath, Anja Naumann, Michael Nischt, Katrin Wolf, Blazej Lewcio, and Maija Poikela for their friendly support, their communicative motivation in their work, and/or the relaxing time spent playing table football.

Finally, and obviously, I deeply thank Odile, my sisters, my parents, my family in general, and my friends (to which many of the above-mentioned colleagues belong) for their love and support.

Contents

Acronyms

AAC	Advanced Audio Coding
AAC-LC	Advanced Audio Coding Low Complexity
AC-3	Dolby AC-3
ACR	Absolute Category Rating
AoS	Acceptability of Service
ARQ	Automatic Repeat reQuest
CABAC	Context-Adaptive Binary Arithmetic Coding
CBR	Constant BitRate
CRT	Cathode Ray Tube
DCR	Degradation Category Rating
DCT	Discrete Cosine Transform
DSCQS	Double Stimulus Continuous Quality Scale
DSIS	Double Stimulus Impairment Scale
EBU	European Broadcasting Union
ES	Elementary Stream
FEC	Forward Error Correction
FFT	Fast Fourier Transform
FR	Full Reference
GOP	Group of Picture
HD	High Definition
HE-AAC	High Efficiency Advanced Audio Coding
HVS	Human Visual System
IDR	Instantaneous Decoder Refresh
IP	Internet Protocol
IPTV	Internet Protocol TeleVision
ITU	International Telecommunication Union
LCD	Liquid Crystal Display
MDCT	Modified Discrete Cosine Transform
MDS	Multidimensional Scaling
MM	Markov Model
MOS	Mean Opinion Score
MoTV	Mobile TeleVision
MP2	MPEG-1 Layer 2
MP3	MPEG-1 Layer 3

MPEG2-TS	MPEG2 Transport Stream
MSE	Mean Squared Error
MTU	Maximum Transmission Unit
MUSHRA	MUltiple Stimuli with Hidden Reference and Anchor
NAL	Network Adaptation Layer
NR	No-Reference
PC	Paired Comparison
PCA	Principal Component Analysis
PCAP	Packet CAPture
PES	Packetized Elementary Stream
PLC	Packet-Loss Concealment
P.NAMS	Parametric Non-intrusive Assessment of audiovisual Media Streaming quality
P.NBAMS	Parametric Non-intrusive Bitstream Assessment of video Media Streaming quality
P.OLQA	Perceptual Objective Listening Quality Assessment
PPS	Picture Parameter Set
PSNR	Peak-Signal-to-Noise Ratio
QoE	Quality of Experience
QoS	Quality of Service
RLC	Run-Length Coding
RMSE	Root Mean Square Error
RR	Reduced-Reference
RTP	Real-time Transport Protocol
SAMVIQ	Subjective Assessment Method for VIdeo Quality
SD	Standard Definition
SMR	Signal to Masker Ratio
SPS	Sequence Parameter Set
SSCQE	Single Stimulus Continuous Quality Evaluation
STB	Set Top Box
TS	Transport Stream
TV	TeleVision
UDP	User Datagram Protocol
VBR	Variable BitRate
VLC	Variable-Length Coding
VoD	Video on Demand

Chapter 1
Introduction

1.1 Motivation

In order to achieve a high degree of user satisfaction for current and upcoming video services like Video on Demand (VoD), Internet Protocol TeleVision (IPTV), and Mobile TeleVision (MoTV), perceived quality needs to be estimated both in the network planning phase—when the network is not yet set-up—and as part of the service monitoring—once the network is set-up and running. Quality assessment can be achieved using audiovisual quality tests or by instrumental methods, which yield estimates of audiovisual quality as perceived by the user. If properly conducted, quality tests with human subjects are the most valid way to assess quality, since quality results from a perception and interaction process in the mind of the user. However, since subjective tests are time consuming, costly, and do not allow to assess the quality during real-time service operation, instrumental assessment methods are often preferred.

1.2 Objectives and Scope

The objective of this work is to provide a *parametric packet-based* audiovisual quality model for IPTV services. The model is to be used both for network planning and quality monitoring in the case of encrypted data. The addressed transport protocol is MPEG2 Transport Stream (MPEG2-TS) over RTP over IP.

The audiovisual quality model will be composed of three modules for the respective audio, video, and audiovisual components. The same version of the audio quality model will be used for both network planning and quality monitoring, while different versions of the video quality model will be used for the two application types. The audio and video quality models take as input a parametric description of the processing and transmission path, from the encoder to the decoder, and work on PES- and TS-encrypted data. Parameters are extracted from packet-header

M.-N. Garcia, *Parametric Packet-Based Audiovisual Quality Model for IPTV Services*,
T-Labs Series in Telecommunication Services, DOI: 10.1007/978-3-319-04855-0_1,

information, the packets being acquired from a network interface or from network traces, typically in the format of Packet CAPture files. As output, the models provide individual estimates of the audio, video, and audiovisual quality. The typical duration of the measurement window is 10 s.

The models do not require access to the original (un-degraded) signal and to the decoded signal, but are provided with side-information such as the employed codec and packet-loss-concealment.

The audiovisual model covers typical audio and video degradations for IPTV and addresses SD (Standard Definition), HD1080 (High Definition, $1{,}920 \times 1{,}080$ pixels) and HD720 (High Definition, $1{,}280 \times 720$ pixels) video formats. The model supports the H.264 video codec and the audio codecs MPEG-1 Layer 2 (MP2), MPEG-1 Layer 3 (MP3), Advanced Audio Coding (AAC), High Efficiency Advanced Audio Coding (HE-AAC) and Dolby AC-3 (AC-3). In terms of IP-network layer transmission errors, the models handle both random and bursty losses.

1.3 Results

Since the models should estimate the quality as perceived by the user, they were developed based on the results of numerous subjective tests. Three sets of databases are described in this work: the “TVM”, the “P.NAMS training” and the “P.NAMS validation” database sets (“P.NAMS” stands for “Parametric non-intrusive assessment of audiovisual media streaming quality”). The TVM set was entirely developed within the “T-Labs” (Telekom Innovation Laboratories) T-V-Model (TVM) project, which was partially funding this work. The two P.NAMS database sets were produced under NDA between the ITU-T P.NAMS and P.NBAMS participants during the respective ITU-T competition in the workgroup Question Q.14/12.

The TVM set and the P.NAMS training set were also extensively used in this work for analysing the quality impact due to typical IPTV degradations, mainly audio and video compression artifacts and transmission errors, studying the relation between subjective test results and technical transmission characteristics as well as audio and video content properties. The relation between audio quality and video quality and their impact on the audiovisual quality were also carefully studied when developing the audiovisual quality model. Finally, the quality impact due to the audio and video contents was analysed at each modeling step.

A first version of the audio, video, and audiovisual quality models was developed in the context of the TVM project. The models addressing the quality monitoring application served as the basis for the models submitted to the ITU-T P.NAMS competition for the Higher Resolution application area (IPTV, VoD). The resulting models show high performance results—in terms of R: Pearson Correlation Coefficient, and *RMSE*: Root-Mean-Square-Error—with unknown data. All performance results are summarized in Table 1.1.

Table 1.1 Performance results for all proposed models in this work

Model	*R*	*RMSE*
Audio	0.92	0.39
Video (NP)	0.84	0.56
Video (QM)	0.91	0.44
Audiovisual (NP)	0.86	0.52
Audiovisual (QM)	0.92	0.42

NP network planning; *QM* quality monitoring; *R* Pearson correlation coefficient; *RMSE* root mean square error (on a 5-point scale)

1.4 Contribution by the Author

The author's contribution to the scientific domain in relation to this work is twofold: the acquisition and analysis of subjective test results for IPTV-type audiovisual streaming media, and the development of the audio, video, and audiovisual quality models for IPTV services.

Tasks related to the acquisition of subjective test results include the selection of appropriate test methods and rating scales, and the design of the tests. The analysis of the subjective tests resulted in a deeper understanding of the link between audiovisual degradations and the perceived quality, and the interaction between audio quality and video quality in the context of IPTV services. Also provided and resulting from an extensive review of the literature is a link between the perceived quality and the whole Quality of Experience (QoE).

All models were developed based on the subjective test results. As part of this work, the author has provided a method for merging subjective test results from different tests. Based on the analysis of the subjective test results and of the audio and video degradations, the author has identified, extracted and computed parameters to be used as input to the models, and selected the most appropriate function for mapping the parameters to the perceived quality. Once the parameters and mapping functions selected, the models have been trained on all available data. For the submission to the ITU-T P.1201 (ex P.NAMS) competition, the robustness of the model was tested using cross-validation. After the model submissions, new databases were created by the competition group, which were used for model validation.

The scientific contribution of this work has been published in a variety of conference and journal papers, book chapters, patent applications, ITU contributions, and finally the ITU-T P.1201 and ITU-T P.1201.2 standards. These publications have served as the basis for the present work.

Conference Papers

- A. Raake, M.-N. Garcia, S. Möller, J. Berger, F. Kling, P. List, J. Johann and C. Heidemann (2008). T-V-Model: Parameter-based Prediction of IPTV Quality. Proceedings of the IEEE International Conference on Acoustics, Speech and Signal Processing (ICASSP).

- M.-N. Garcia, A. Raake and P. List (2008). Towards content-related features for parametric video quality prediction of IPTV services. Proceedings of the IEEE International Conference on Acoustics, Speech and Signal Processing (ICASSP).
- B. Feiten, A. Raake, M.-N. Garcia, U. Wüstenhagen and J. Kroll (2009). Subjective quality evaluation of audio streaming applications on absolute and paired rating scales. Proceedings of the 126th Audio Engineering Society (AES) convention.
- M.-N. Garcia and A. Raake (2009). Impairment-factor-based audio-visual quality model for IPTV. Proceedings of the first International Workshop on Quality of Multimedia Experience (QoMEX).
- M.-N. Garcia and A. Raake (2010). Quality impact of audio on audio-visual quality in the context of IPTV. Fortschritte der Akustik (DAGA10), 36 Jahrestagung für Akustik. Deutsche Gesellschaft für Akustik e.V. (DEGA).
- M.-N. Garcia and A. Raake (2010). Quality impact of video format and scaling in the context of IPTV. Proceedings of the third International Workshop on Perceptual Quality of Systems (PQS).
- M.-N. Garcia, R. Schleicher and A. Raake (2010). Towards a content-based parametric video quality model for IPTV. Proceedings of the fifth International Workshop on Video Processing and Quality Metrics for Consumer Electronics (VPQM).
- M.-N. Garcia and A. Raake (2010). Parametric packet-layer video quality model for IPTV. Proceedings of the tenth International Conference on Information Science, Signal Processing and their Applications (ISSPA).
- M.-N. Garcia and A. Raake (2010). Normalization of subjective video test results using a reference test and anchor conditions for efficient model development. Proceedings of the second International Workshop on Quality of Multimedia Experience (QoMEX).
- M.-N. Garcia and A. Raake (2011). Frame-layer packet-based parametric video quality model for encrypted video in IPTV services. Proceedings of the third International Workshop on Quality of Multimedia Experience (QoMEX).
- M.-N. Garcia, A. Raake and B. Feiten (2013). Parametric audio quality model for IPTV services—ITU-T P.1201.2 audio. Proceedings of the fith International Workshop on Quality of Multimedia Experience (QoMEX).
- M.-N. Garcia, P. List, S. Argyropoulos, D. Lindegren, M. Pettersson, B. Feiten, J. Gustafsson and A. Raake (2013). Parametric model for audiovisual quality assessment in IPTV: ITU-T Rec. P.1201.2. Proceedings of the 15th IEEE International Workshop on Multimedia Signal Processing (MMSP).

Journal Papers

- M.-N. Garcia, R. Schleicher and A. Raake (2011). Impairment-factor-based audiovisual quality model for IPTV: Influence of video resolution, degradation type, and content type. EURASIP Journal on Image and Video Processing, vol. 2011, pp. 1–14.
- Q. Huynh-Thu, M.-N. Garcia, F. Speranza, P.J. Corriveau and A. Raake (2011). Study of rating scales for subjective quality assessment of high-definition video. IEEE Trans. on Broadcasting, vol. 57, no. 1, pp. 1–14.

- A. Raake, J. Gustafsson, S. Argyropoulos, M.-N. Garcia, D. Lindegren, G. Heikkilä, M. Pettersson, P. List and B. Feiten (2011). IP-based mobile and fixed network audiovisual media services. IEEE Signal Processing Magazine, vol. 28, pp. 68–79.

Book Chapters

- M.-N. Garcia, S. Argyropoulos, N. Staelens, M. Naccari, M. Rios-Quintero and A. Raake (to be published in 2014). Chapter 19: Video Streaming. Quality of Experience: Advanced Concepts, Applications and Methods. Edited by S. Möller and A. Raake, Springer.
- B. Feiten, M.-N. Garcia, P. Svensson and A. Raake (to be published in 2014). Chapter 16: Audio transmission. Quality of Experience: Advanced Concepts, Applications and Methods. Edited by S. Möller and A. Raake, Springer.

Patents

- M.-N. Garcia, S. Möller and A. Raake. US2009244289: Audio-visual quality estimation.
- M.-N. Garcia, B. Feiten, S. Argyropoulos, P. List and A. Raake. WO2012013777: Method and apparatus for assessing the quality of a video signal during encoding or compressing of the video signal.
- A. Raake, M.-N. Garcia, B. Feiten, S. Argyropoulos and P. List. WO2012076202: Method and apparatus for assessing the quality of a video signal during encoding and transmission of the video signal.
- M.-N. Garcia, B. Feiten, S. Argyropoulos, P. List and A. Raake. EP2525353: Parametric audio quality model for IPTV services.
- M.-N. Garcia, S. Argyropoulos, P. List, B. Feiten, and A. Raake. Content-dependent video quality model for video streaming services.

ITU Contributions

The following ITU contributions were written in the course of the ITU-T SG12 Q14 P.NAMS competition. Some of these contributions, such as the proposal for subjective databases, result from a collaborative effort between P.NAMS participants. P.NAMS subjective training and validation databases were eventually designed and produced by all P.NAMS participants. Similarly, the ITU-T P.1201 recommendation results from a joint effort of the ITU-T SG12 Q14 participants. The ITU-T P.1201.2 Recommendation was conjointly written by the authors of this work with P. List, and S. Argyropoulos, from T-Labs. It was eventually reviewed by the ITU-T SG12 Q14 participants.

- A. Raake, M.-N. Garcia and S. Möller (2008). DTAG View for P.NAMS and G.OMVS standardization procedure. International Telecommunication Union.
- M.-N. Garcia, B. Feiten and A. Raake (2010). Proposed changes of P.NAMS ToR for the IPTV application area. International Telecommunication Union.
- M.-N. Garcia, D. Lindegren and S. Broom (2010). Subjective databases required for P.NAMS standardisation. International Telecommunication Union.

- M.-N. Garcia, B. Feiten and A. Raake (2010). Subjective quality evaluation of audio streaming applications: ACR versus MUSHRA. International Telecommunication Union.
- M.-N. Garcia and A. Raake (2011). P.NAMS/P.NBAMS Training Database—Proposal for common set. International Telecommunication Union.
- M.-N. Garcia, S. Argyropoulos, S. Broom and A. Raake (2011). P.NAMS/P.NBAMS Training Database—Proposal for the number of tests per database and the selection of test conditions. International Telecommunication Union.
- ITU-T P.1201 (2012): Parametric non-intrusive assessment of audiovisual media streaming quality. International Telecommunication Union.
- ITU-T P.1201.2 (2012): Parametric non-intrusive assessment of audiovisual media streaming quality—higher resolution application area.

Contributions from T-Labs' Colleagues

During the development of the P.1201 and P.1201.2 standards, some tasks were conducted by colleagues of the author of this work, and hence are not covered by this work:

(1) Data acquisition from the PES- and TS-encrypted stream,
(2) video scene-cut detection,
(3) extension of the model to unencrypted data.

The interested reader is referred to the following conference papers and patents, as well as to the ITU-T P.1201.2 Recommendation ITU-T Recommendation G.107 (2005).

Conference Papers

- S. Argyropoulos, A. Raake, M.-N. Garcia and P. List (2011). No-reference video quality assessment of SD and HD H.264/AVC sequences based on continuous estimates of packet loss visibility. Proceedings of the third International Workshop on Quality of Multimedia Experience (QoMEX).
- S. Argyropoulos, A. Raake, M.-N. Garcia and P. List (2011). No-Reference bit stream model for video quality assessment of H.264/AVC video based on packet loss visibility. Proceedings of the IEEE International conference on Acoustics, Speech, and Signal Processing (ICASSP).

Patents

- S. Argyropoulos, B. Feiten, M.-N. Garcia, P. List and A. Raake. WO2012076203: Method and apparatus for objective video quality assessment based on continuous estimates of packet loss visibility.
- B. Feiten, A. Raake, S. Argyropoulos, M.-N. Garcia, and P. List. WO2012034975: Method of and system for measuring quality of audio and video bit stream transmissions over a transmission chain.
- P. List, B. Feiten, S. Argyropoulos, M.-N. Garcia and A. Raake. WO2012013655: Method for estimating the type of the group of picture structure of a plurality of video frames in a video stream.

1.5 Structure of the Book

Chapter 2 provides background information needed for developing and validating the audiovisual quality model. This includes (1) providing definitions of quality and quality-related terms, (2) analysing which elements are impacting quality in the context of multimedia applications, describing the quality impact in terms of perceptual features caused by the service elements, and linking the quality features to the perceived quality, (3) linking quality to Quality of Experience QoE, (4) reviewing existing test methods and rating scales to be used when conducting subjective tests, highlighting which biases may be introduced by these methods and scales, and (5) providing an overview of quality model types, modeling strategies, and model evaluation metrics. Chapter 3 revisits all these steps but focusing this time on the one-way IPTV services considered by the models developed in this work. The model scope and type will be described in detail, quality elements and features will be re-addressed. All subjective tests conducted for developing the models will be presented, too. This includes a description of the used contents, test conditions, processing chains, test methods and rating scales. Finally, the selected modeling method will be presented as well as the procedure for evaluating the model. The results of the audio, video, and audiovisual subjective tests will be extensively analysed in Chaps. 4–6, respectively. Following these analyses, the models will be presented along with the selected input parameters, then evaluated. One single audio quality model will be presented in Chap. 4, which will be used for both network planning and quality monitoring application. In turn, two video quality models have been developed and are presented in Chap. 5, one addressing the network planning application, the other model the quality monitoring case. Finally, two modeling approaches will be tested for the audiovisual quality models in Chap. 6, resulting in two models for network planning, and two models for quality monitoring. All model variants are tested and compared. Finally, Chap. 7 provides a discussion of and an outlook on the conducted and future work.

Chapter 2
Quality Engineering

The goal of this work is to develop an audiovisual quality model for estimating the quality as perceived by the user of video streaming services such as IPTV. It aims at presenting the steps to be completed for fulfilling this objective. It also covers background information necessary for understanding the choices made at each step of the model development and puts the model in a general Quality of Experience (QoE) framework. The structure of this chapter follows the chart of Fig. 2.1.

This chapter will therefore start with a presentation of the concepts of quality, Quality of Service (QoS) and Quality of Experience (QoE). The key concepts of quality elements and quality features (Block "Define" in Fig. 2.1) will also be defined. These definitions allow a better design of the subjective quality tests to be conducted for developing the quality model. They also help in selecting which modeling strategy should be followed for obtaining the best performing model. Lastly, the concepts of QoS and QoE allow understanding how the developed quality model should be used and combined with other metrics, depending on the application, for ensuring a high user satisfaction.

The second section details quality elements and features associated with the IPTV application. Broadly speaking, quality elements are the technical and/or design elements which eventually have an impact on quality. Quality features correspond to the description of the perceptual impact of the signal characteristics and degradations. Quality elements belong to the physical domain while quality features belong to the perceptual domain (Jekosch 2005). As perceived quality is by nature multidimensional, perceptual quality dimensions—defining the sub-dimensional space underlying the quality features—are discussed to link the quality features to the perceived quality. The identification of the quality features determine which signal degradations should be shown to the subjects during the tests. The link between quality elements and quality features should also be understood. Appropriate values can then be given to the quality elements for obtaining the targeted degradation types and strengths in the subjective tests.

The third and fourth sections of this chapter give an overview of existing test methods, rating scales and modeling development strategies found in the literature in the

M.-N. Garcia, *Parametric Packet-Based Audiovisual Quality Model for IPTV Services*, T-Labs Series in Telecommunication Services, DOI: 10.1007/978-3-319-04855-0_2,

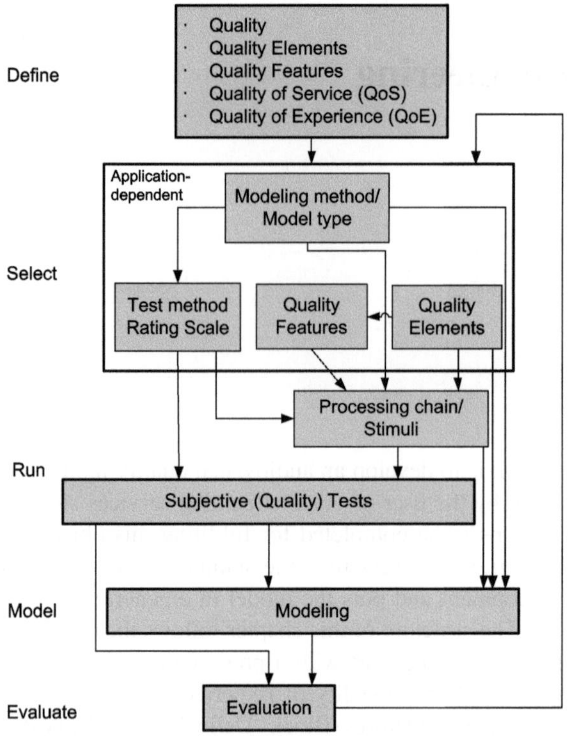

Fig. 2.1 Quality engineering flow chart

context of audiovisual quality assessment and modeling (block "Select" in Fig. 2.1). The objective here is not to provide an exhaustive overview of test methods and rating scales. Instead, the goal is to present respective problems and possible solutions associated with the choice of test method, rating scale, modeling methodology and evaluation.

The actual selections made for the IPTV application studied in this work are presented in Chap. 3. The audio, video, and audiovisual subjective quality tests and the models developed based on the results of these tests are presented in Chaps. 4, 5 and 6 (Blocks "Run" and "Model" in Fig. 2.1). Model performance indicators (Block "Evaluate" in Fig. 2.1) are also provided.

2.1 The Concept of Quality

This section starts with some definitions related to quality provided by Jekosch in Jekosch (2005). Möller (2000) has complemented her work by a taxonomy of QoS and acceptability in the context of telephone services. A comparison to other taxonomies found in the literature, especially in the literature related to image and video quality, will be provided.

2.1.1 The Jekosch School

Quality has been defined by Jekosch in (2005) as

> [the] result of [the] judgment of the perceived composition of an entity with respect to its desired composition.

The *perceived composition* is the

> [t]otality of features of an entity,

and a *feature* is

> [a] recognizable and nameable characteristic of an entity.

Lastly, the *desired composition* is

> [t]he totality of features of individual expectations and/or relevant demands and/or social requirements.

These definitions have been developed in the context of voice and speech quality assessment. They are based on the ISO-9000 standard (ISO 2000–2012) and its supplements (DIN 2008–2005a, b), where the definitions of quality and quality-related terms are not restricted to speech. Jekosch's definitions are valid for any kind of modalities and respective quality perception, for example audio quality, video quality, food quality, perfume quality, and even services or products that involve multiple modalities such as in the case of audiovisual quality. The above definition will be applied throughout this work to audio (including speech), video and audiovisual quality. The "entity" is therefore either a sound, an image (a video) or the combination of both. The "perceived compositions" are then perceived auditory, visual, and audiovisual compositions. Visual features are, for instance, brightness, form, color, and motion information (Jumisko-Pyykkö 2007) whereas pitch, loudness, timbre, duration and location/spaciousness (Raake 2006b; Jumisko-Pyykkö 2007) are relevant auditory features.

These definitions highlight several important aspects of quality:

- Quality is seen from the perspective of the observer.
- Quality is the result of both a perception process and an assessment process.
- The perceptual event resulting from the signal event under observation is composed of various "features" and is therefore of a multidimensional nature.
- The observer has an "internal reference" for each feature against which she/he compares the perceived feature.

For a better understanding of the term "quality", the scheme created by Raake in (2006b) is analysed. This scheme combines the concepts by Blauert and Jekosch (1997) and Jekosch (2005). It represents a test subject in a listening quality test. It is shown in Fig. 2.2 with additional legends, for clarification purpose. The attention of the user has also been added for representing its impact on the perception process, as described below. Finally, since they are compared to the desired quality features,

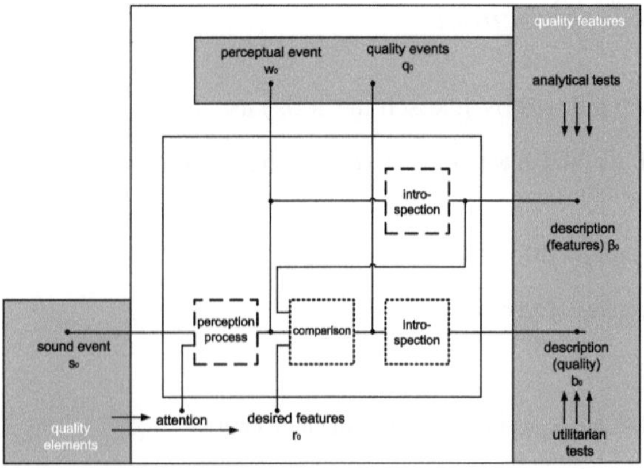

Fig. 2.2 Listener in a quality judgement context, adapted from Raake (2006b)

the observed quality features have been added as input to the "comparison" box. Explanations are given in the following.

The left bold dashed square corresponds to the perception process, which transforms, in this case, a sound event, i.e., the signal the listener perceives, into a perceptual (auditory) event. The perceptual event cannot directly be measured, since it is internal to the observer. It can however be assessed by asking the listener to describe it based on introspection: this step is represented by the right bold dashed square. At this stage, the listener did not assess the quality but provides a description of the features of the perceived signal (sound event). Note that those features are however related to quality, like noisiness or loudness, and are named "quality features". More precisely, a *quality feature* is Jekosch (2005)

> [a] recognized and designated characteristic of an entity that is relevant to the entity's quality.

During the assessment process—which is represented by non-bold dash squares in Fig. 2.2—each perceived feature is compared to its desired corresponding feature, resulting in a quality event. As for the perceptual event, the quality event can be measured only indirectly by asking the observer to assess (describe) the quality.

Note that the "desired features" can be considered as the "features of the" *internal reference* which was introduced earlier in this section. Figure 2.3 gives a schematic view of factors impacting the *internal reference* and the *attention* of the user. The *internal reference* integrates aspects influencing the expectation of the user, and therefore the desired features of the user.

One of these aspects is the previous *experience* the user has with the addressed technology/service.

Also modifying the *internal reference* of the user, and therefore impacting the perceived quality, is the *context* in which the technology/service is assessed.

Context may refer to the *environmental* context, i.e., the situation and location of the user. This is of particular importance for mobile applications such as MoTV (Mobile TV), for which the perceived quality may depend on the location where the

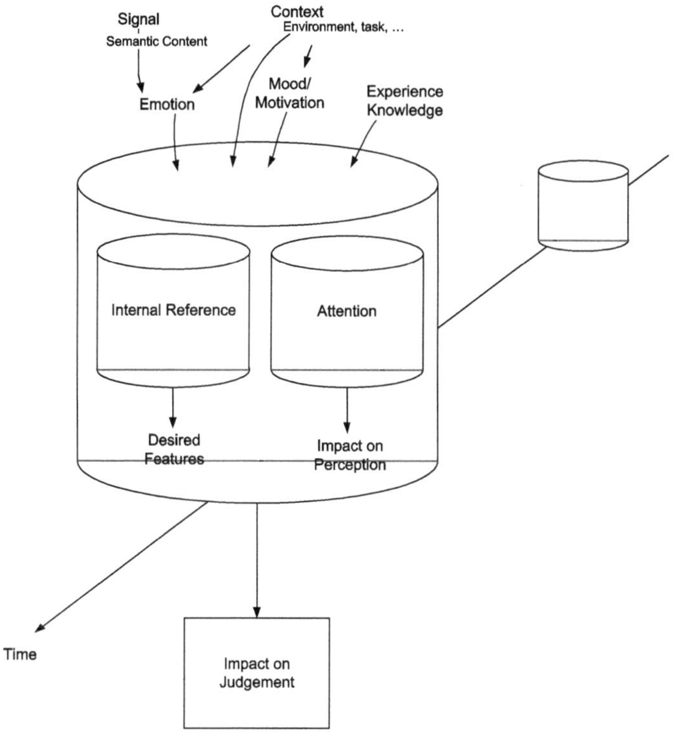

Fig. 2.3 User-related factors impacting quality judgements

user is watching the video, e.g., at home versus in a cafe or in a bus (Jumisko-Pyykkö and Hannuksela 2008). Note that, as is shown in Gros et al. (2004) for time-varying speech, the context does not necessarily have an influence on the perceived quality.

Context also refers to the situation in which the user is, e.g., making an emergency call vs. a non-important call. It also includes economical aspects such as the price paid by the user for getting the service, and social aspects, i.e., the amount and type of persons involved in the experience (Jumisko-Pyykkö 2011; Le Callet et al. 2012). These contextual aspects may impact the *mood* or the *emotional state* in which the user is.

This emotional state may also be influenced by the *semantic content* of the signal.

The above factors are present whether an assessment *task* is given to the user or not. In case the user has to accomplish an assessment *task*, her/his *internal reference* is modified by this *task*, in particular by the selected assessment method, and the range of stimuli presented to the subjects. Those latter aspects will be covered in Sects. 2.3.1 and 2.3.2.

Some of the factors influencing the *internal reference*, such as the previous *experience* of the observer with the assessed item or the *task* given to the observer, are to some extents also impacting the perception process. For instance, an expert and a non-expert in video encoding may focus their *attention* on different parts of the video. The *task* given to the observer or her/his emotional state may also direct her/his *attention* to different characteristics of the assessed item. These aspects are also considered in Neisser's perception model (Neisser 1976), which is described in Sect. 2.1.2.

The *internal reference* and the *attention* of the user are therefore influenced by many factors. Since these factors are time-varying, the *internal reference*, the *attention* and therefore the quality judgement are time-varying, too. This is reflected in Fig. 2.3 by the respective time-axis.

Strongly linked to the notion of *quality features* are the *quality elements*.

A *quality element* is Jekosch (2005)

> [a] [c]ontribution to the quality
>
> - of an immaterial or a material product as the results of an action/activity or a process in one of the planning, execution or usage phases.
> - of an action or of a process as the result of an element in the course of this action or process.

The notion of *quality elements* therefore reflects the perspective of the network provider or manufacturer of service components and system. As such, quality elements consider the impact of the design and service planning implementation and operation on quality.

There is not a one-to-one relation between *quality elements* and *quality features*: several *quality elements* can be linked to one *quality feature*, and conversely one *quality element* can be linked to several *quality features*. For instance, the combination of the video resolution (number of pixels per video frame), the employed codec and the selected bitrate impact the blockiness[1] of the picture, but they also influence the quality impact due to packet loss.

The notions of *quality element* and *quality feature* will be extensively used throughout this work. More information on *quality elements* and *quality features* and how they relate to each other will be provided in Sect. 2.2. Since the target application is IPTV, focus will be given on the quality elements and the quality features related to audio, video, and audio-video in the context of IPTV.

2.1.2 Comparison with Other Terminology in the Literature

It should be noted that the concepts related to quality and presented in the previous section are stemming from studies targeting speech quality assessment in the context of telecommunications. As previously mentioned, and since they are partly adapted from standards with wider scope (ISO 2000–2012), the proposed definitions are not specific to speech, and can be applied to other individual modalities. Since a wide

[1] Blockiness—or block distortion—is defined in ITU P.930 as the distortion of the image characterized by the appearance of an underlying block encoding structure.

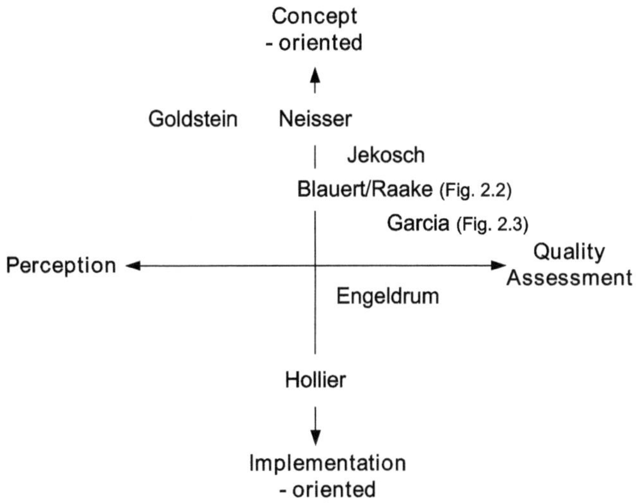

Fig. 2.4 Classification of perceptual quality theories

portion of the work presented in this book is dedicated to video quality assessment, it is of interest to compare Jekosch's quality-related concepts to other theories from the literature, in particular from the image or video quality domain. In addition, this section addresses the validity or extension of the concepts when applied to multiple modalities such as in the case of audiovisual signals.

Note that the theories presented in the following may differ because they have not been developed for exactly the same purpose.

Figure 2.4 is an attempt to capture the different objectives of the proposed theories. For instance, Goldstein's work (Goldstein 2002), on the left side of the figure, targets the "description" ("concept") of the "perception" process, while Jekosch ultimately "describes" the "quality assessment" process. On the other hand, these two theories focus on the "description" (top of the figure) of the perception and quality assessment processes, while works from, e.g., Engeldrum (2000) and Hollier et al. (1999) (bottom of the figure) investigate the modeling steps ("implementation") to be undertaken for obtaining high-performance perceptual quality models.

Goldstein (2002) often cited in the audiovisual domain, splits the human perception processing in two levels: the low-level sensorial and the high-level cognitive processing levels. Broadly speaking, the low-level sensorial level corresponds to all processing steps from the reception of the (auditive, visual, olfactory...) signal to an understandable form for the neural processes of the brain. The high-level cognitive processing level interprets this transformed signal and allows decision making such as the formation of the judgment. In terms of Jekosch's terminology, the low-level sensorial level would correspond to the "perception process" box (see Fig. 2.2) while the high-level cognitive processing level corresponds to the "introspection" and "comparison" boxes (see Fig. 2.2).

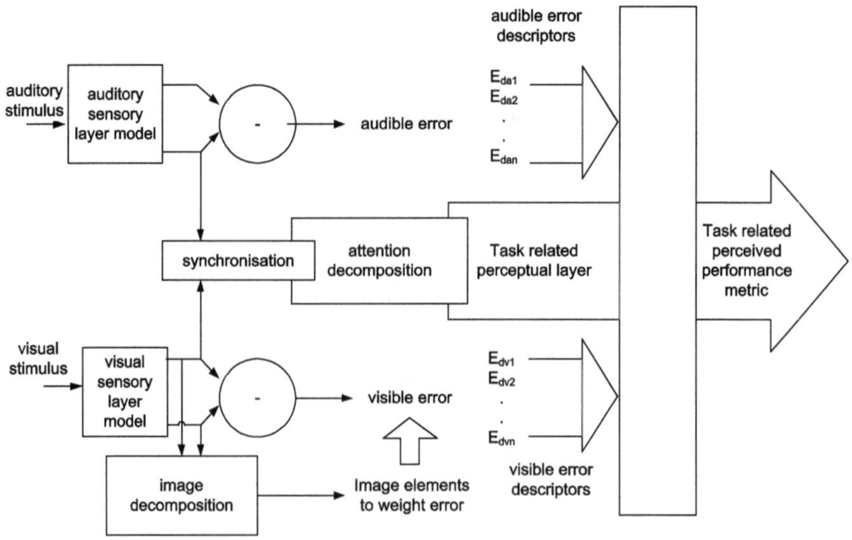

Fig. 2.5 Hollier's multisensory perceptual model [reproduced from Hollier et al. (1999)]

This processing-level approach can also be seen in Hollier et al.'s multisensory perceptual model (Hollier et al. 1999) shown in Fig. 2.5. This model addresses multi-modal signals and is composed of two layers: the sensory layer and the perceptual layer. They correspond to the low-level sensorial processing level in Goldstein's approach, and prepare the received sensorial signal for the cognitive processes, and therefore the judgement and decision making processes. The sensory layer reproduces the properties of our senses, e.g., hearing and seeing. The perceptual layer is already at the border between sensorial and cognitive processes. In Hollier's terms, the perceptual layer attempts to predict the subjective importance of perceptive (e.g., auditory domain) errors. Indeed, when receiving a complex sensorial signal, all characteristics of the signal are not perceived equally, with some characteristics impacting—increasing or masking—the perception of other characteristics. For instance, high-frequency noise inserted in a highly textured image will not be perceived, while it will be easily detected in a content with low amount of texture.

At this stage, the border between sensorial and cognitive processes is not so distinct anymore. Indeed, and as previously mentioned, the attention of the person is impacting the perception (audibility, visibility) of the signal, this attention being itself driven by the presence or not of additional modalities (multi-modal perception), the motivation and experience of the person and the task she/he has to complete.

According to Jekosch's terminology, the "sensory layers" correspond to the perception process, the "audible and visible errors" to the "perceptual events", and the "audible and visible error descriptors" are mapped to the "quality features" in the "task related perceptual layer".

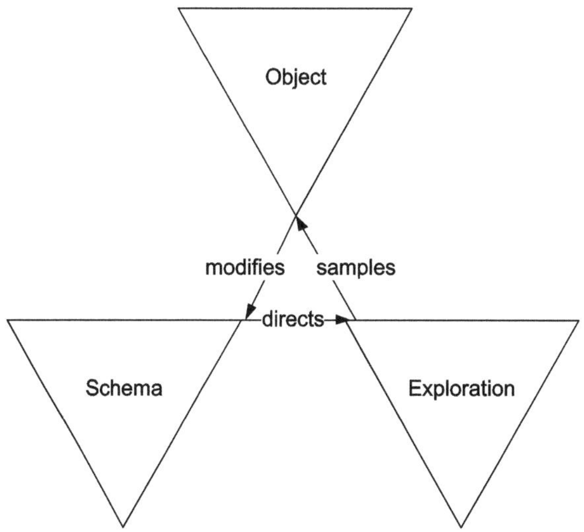

Fig. 2.6 Neisser's Perceptual Cycle [reproduced from Neisser (1976)]

It should be noted that Hollier et al. further detail the decision-making process (Hollier et al. 1999). They describe the cognitive components—based on Wickens' work (Wickens 1992)—such as the long-term memory, the working memory, the attentional processing and the current goal and task undertaken by the person. Some important detailed effects impacting the memory, such as the recency and primary effects, or the visualized content in the case of visual perception, are also considered. Indeed, due to human memory limitations, people tend to remember best what comes first ("primary effect"), and what comes last ("recency effect") (Aldridge et al. 1998; Hands and Avons 2001). These effects should be kept in mind when designing subjective tests: if a stimulus contains temporal degradations, degradations coming at the beginning or at the end of the stimulus will be better remembered by the observers. These effects are commonly reduced by using short-duration (10 s) stimuli in subjective tests. The interested reader is referred to Hollier et al. (1999) for further information.

Also noteworthy in this context is Neisser's model (Neisser 1976) introduced in the previous section and displayed in Fig. 2.6.

In this perceptual cycle, "Object" represents the information available to human senses. A "schema" is defined as (Neisser 1976)

> that portion of the perceptual cycle that is inside the observer, modifiable by experience, and somehow specific to what is being observed [...].

Note that the term "schema" is referred to as "schemata" by Piaget et al. in Piaget (1924, 1937). Jekosch mentions Piaget's work as well in Jekosch (2005, Sect. 5.2). In her terminology, the "schema" is also referred to as "internal reference". For Neisser, it represents one's knowledge about the stimulus environment, and is based on previous experiences. The "schema" *directs* the attention, it therefore impacts the perception process. Different *samples* are selected from the "Object" to be judged, and the resulting "perceptual event" in Jekosch's terminology is modified. If the selected stimuli do not fit the expectations laid by the schema, the schema may be modified. This process is called "accommodation" in Piaget's terminology. According to Piaget, there is an "assimilation" strategy in addition to the "accommodation". In that case, the perceived signal does not fulfill the expectation and is aligned with the existing "schema". Neisser does not see the need for including this strategy in the perceptual cycle.

The Neisser's model highlights that perception is a continuous process, and how sensitive the "schema"—or "internal references" are. It varies over time for a given subject, but also between subjects, where the subject is the observer and judge in the case of quality assessment. The features against which the subject is comparing the observed features may continually change. This aspect should not be underestimated when conducting subjective quality tests, especially when addressing applications the subjects are not familiar with. For applications such as telephony or TV, the schema can be considered as more or less hard-coded in human's brains, and therefore the "schemas" can be considered as stable. However, when conducting subjective tests, and whatever the application is, it is recommended to design a training session in which subjects will be presented the same quality range and degradation types than in the actual test session. This should help in stabilizing the "schema".

The work by Engeldrum (2000) is frequently cited in image or video studies (Heynderickx and Muijs 2005). He uses the following definition of quality (Engeldrum 2004):

> *Quality* is the integrated perception of the overall degree of excellence of the image.

The term "integrated" refers to the combination of the quality features of the image—the "-nesses", or "Customer Perceptions" (perceptual attributes) in the Engeldrum's Image Quality Circle model (see Fig. 2.7).

This model relates measurable technological variables to the image quality and is composed of four blocks: the "Technology Variables", the "Physical Image Parameters", the "Customer Perceptions—The "Nesses"", and the "Customer Image Quality Rating". The "Technology Variables" correspond to the "quality elements" presented in the previous section. In the audio domain, they are, for instance, the used audio codec type and profile, or the number of channels. In the video domain, they are, for instance, the video resolution and the employed video codec type and profile. The "Physical Image Parameters" are metrics corresponding to various "Technology Variables". The "Physical Image Parameters" do not really have an equivalence in Jekosch's terminology. This difference comes from the fact that Engeldrum's model intends to describe all modeling steps to be followed for obtaining a high-performance model, while Jekosch's model describes the perception and judgement formation process.

In summary, the concepts of quality and models of quality assessment are not differing much from one study to another one. Parallels between the different employed terminologies can be found. Some studies put more emphasis on the sensory-layer process while other studies describe the cognitive processes more in detail. This

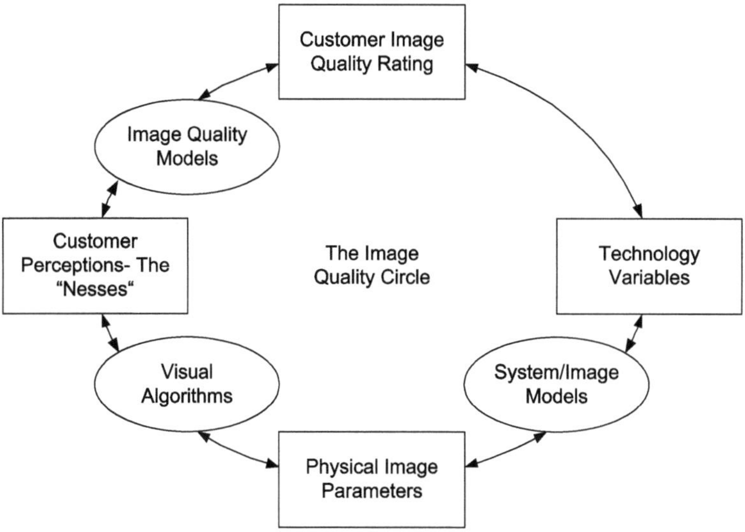

Fig. 2.7 Engeldrum's Image Quality Circle model [reproduced from Engeldrum (2004)]

depends on the perspective taken and on the purpose of the quality model: When designing and interpreting the subjective quality tests, factors impacting the decision-making process such as the undertaken task or the experience of the subject should be known; signal-based quality metrics have been improved with the increase of knowledge on the sensory-layer (e.g., ITU-T Recommendation P. 863); for complex signals yielding several perceptual/quality features, the knowledge of sensory-layer information is not sufficient, and the identification and combination of the quality-relevant features cannot be avoided.

Most studies are generally describing the sensory- and cognitive-processes separately, but they also point out that these processes are interacting. How this interaction works is still not completely known.

2.1.3 Quality of Service and Quality of Experience

In the context of IPTV, and more generally of telecommunications, four more concepts still have to be defined: the "overall" quality, the "integral" quality, QoS, and QoE.

Overall quality refers to the end-to-end quality (Möller 2000). In the context of IPTV, overall quality includes the whole transmission system from the audiovisual recording to the user. In the context of a telephone-type transmission, "overall quality" refers to the "mouth-to-ear quality", i.e., from one user to another one. Thus, all components of the systems from the creation of the signal to the perception of the signal are considered. Note that the "overall quality" should be measured at the user (reception) side.

Integral quality refers to the perceptual dimensions underlying the audiovisual quality (Möller 2000). The "integral quality" reflects all such dimensions.

QoS is defined in ITU (2008b) as

> [the] [t]otality of characteristics of a telecommunications service that bear on its ability to satisfy stated and implied needs of the user of the service.

It should be noted that this definition differs from the definition of the 1994 version of ITU (1994), which states that QoS is

> [t]he collective effect of service performance which determines the degree of satisfaction of a user of the service.

Still, this definition, at least how it is usually interpreted in the literature [see Le Callet et al. (2012)], remains related to network performance, e.g., to the amount of packet loss or delay. In the context of telecommunications, improving the QoS is synonymous to finding the most suitable settings for the different quality elements of the network, e.g., finding the most suitable codec and bandwidth. "Suitable" means that the quality elements are fulfilling both the user's cost efficiency and satisfaction criteria. Following the example of bandwidth, the network operator will try to find the minimum bandwidth which yields no or an acceptable amount of artifacts at the end user side.

The perspective of the user is further considered in ITU (2008b) when defining the concept of *QoSE* as [see also Ibarrola (2011)]

> [a] statement expressing the level of quality that customers/users believe they have experienced.

It is further noted in ITU (2008b) that

> QoSE has two main components: quantitative and qualitative. The quantitative component can be influenced by the complete end-to-end system effects (network infrastructure). The qualitative component can be influenced by user expectations, ambient conditions, psychological factors, application context, etc.

QoSE here refers to the components influencing the user judgement in a quality assessment context. The quantitative component of QoSE refers to the quality elements of the end-to-end system which have an impact on the user judgement. The qualitative component of QoSE refers to the factors which make the user judgements vary over time, with the application context, and between subjects.

Based on the previous review, it cannot be concluded how to interpret the definition of QoSE. It seems to be an attempt for adding the user and context dependencies to the notion of QoS. In that case, the concept of QoSE is not different from the one of perceived quality, since the perceived quality is user- and context-dependent, making the QoSE redundant and therefore useless.

In ITU (2008a), QoE is defined as:

> the overall acceptability of an application or service, as perceived subjectively by the end-user,

where

- Quality of experience includes the complete end-to-end system effects (client, terminal, network, services infrastructure, etc.), and
- Overall acceptability may be influenced by user expectations and context.

The Acceptability of Service (AoS) is defined in EURESCOM Project P. 807 Deliverable 1 (1998) as

> [a] multidimensional property of a service, describing how readily a customer will use the service, represented as the ratio of the number of potential users to the quantity of the target group.

With these definitions, AoS can be seen as a result of QoE: the user takes the decision on using or not using the service based on the QoE when using the service. For assessing the acceptability, the user will directly be asked if she/he finds the service acceptable or not, including all factors which may impact the acceptability of the service: its cost, the overall (multimedia) quality, its usability, the pleasure of using it and its utility.

Due to the unclear relation between the Acceptability and the QoE, a better definition of QoE has to be found. This was one of the purposes of the Dagstuhl seminar 12181, resulting in the following working definition of QoE, also reported in Le Callet et al. (2012):

> Quality of Experience (QoE) is the degree of delight or annoyance of the user of an application or service. It results from the fulfillment of his or her expectations with respect to the utility and/or enjoyment of the application or service in the light of the user's personality and current state.

"Utility" refers to the perceived fulfillment of the purpose the service is designed for, and "enjoyment" describes how much pleasure the user derives from using the service.

Still according to Le Callet et al. (2012), "personality" is used in terms of "… those characteristics of [the] a person that account for consistent patterns of feeling, thinking and behaving" (Pervin and John 2001) and "current state" in terms of "situational or temporal changes in the feeling, thinking or behavior of a person" [translated from German from Amelang et al. (2006)].

An exact definition of QoE is required to build a solid base for further analysis. From a service provider's perspective, the identification of the QoE components allow a better diagnosis of the reasons why a user is subscribing to a service, and why she/he decides to maintain her/his subscription. This identification helps in selecting and designing appropriate tests for assessing components of or integral QoE. Also crucial for assessing QoE are the analysis of the factors influencing QoE, where "Influenc[ing] Factor" have been defined in Le Callet et al. (2012) as:

> any characteristic of a user, system, service, application, or context whose actual state or setting may have influence on the Quality of Experience for the user.

Mainly based on the work by Jumisko-Pyykkö in (2011), "Influence factors" (IFs) are further classified into three categories: "Human IFs", "System IFs" and "Context IFs", where "Human IFs" are defined as:

> any variant or invariant property or characteristic of a human user. The characteristic can describe the demographic and socio-economic background, the physical and mental constitution, or the user's emotional state.

"System IFs" refer to:

> properties and characteristics that determine the technically produced quality of an application or service (Jumisko-Pyykkö 2011). They are related to media capture, coding, transmission, storage, rendering, and reproduction/display, as well as to the communication of information itself from content production to user.

"Context IFs" are:

> factors that embrace any situational property to describe the user's environment in terms of physical, temporal, social, economic, task, and technical characteristics (Jumisko-Pyykkö 2011).

The following analysis of IPTV quality will be based on the Dagstuhl definition of QoE (Le Callet et al. 2012).

All these definitions emphasize that quality tests reported in the literature, even though they are following standardized methods [e.g., International Telecommunication Union (ITU) and European Broadcasting Union (EBU)], are actually assessing the perceived quality, and not the QoE in an actual usage context. Note that these tests are of course still necessary, since the perceived quality is one crucial aspect of QoE, especially in the case of IPTV services.

Moreover, since quality models are developed based on subjective quality tests, these models are also not estimating the whole QoE, but rather a certain aspect of QoE. This is reflected in the QoE framework depicted in Fig. 2.8. According to this framework, the audiovisual quality model developed in this book can be used to predict the perceived quality of the IPTV application. For getting a whole picture of QoE, additional QoE components such as the *utility* have to be identified, then assessed and modeled, similarly to the (perceived) *quality* component. When assessing these components, all factors affecting the quality judgement and reported in this chapter should be considered. In particular, for a general quality model not targeting an explicit user-dependence, test methods should be carefully selected for minimizing or compensating for the inter- and intra-user dependency of the internal reference of the users.

Also note that most instrumental models [e.g., PESQ (ITU 2001b), P.OLQA (ITU 2011b), J.144 (ITU 2001a)] predict perceived quality for short samples of 8–16 s duration. Such short timescale may not be appropriate for estimating other QoE components such as "usability" or "enjoyment". For modeling QoE, it therefore remains open if QoE should first be estimated from QoE components sampled at short intervals ($\simeq$10 s) and then integrated over time, or if QoE components should first be integrated over time, and then combined into a QoE estimate.

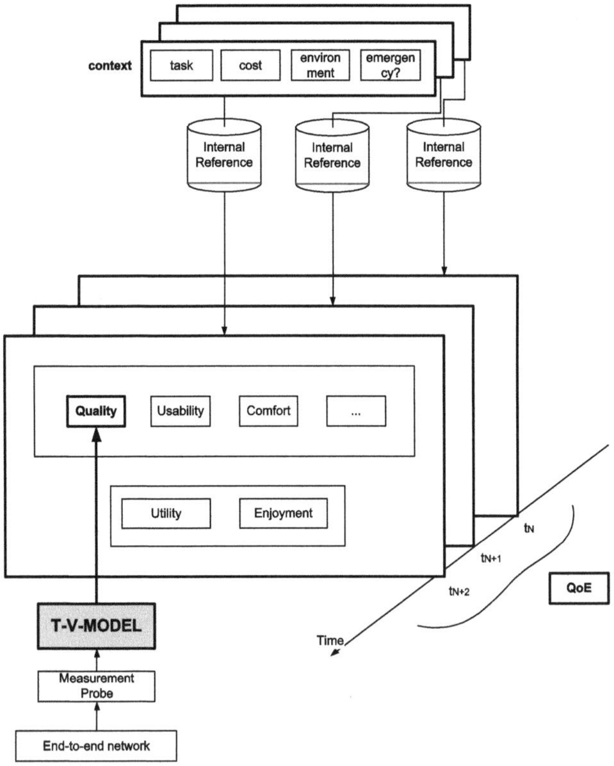

Fig. 2.8 The developed audiovisual quality model (T-V-Model) in the whole QoE framework

For the reader interested in the link between QoS and acceptability, a QoS schematic is proposed in Möller (2000), Sect. 2.2, that was developed for speech communication services. For more details on QoE components and a proposal for linking test methods to the QoE framework, see Geerts et al. (2010).

2.2 From Quality Elements to Quality

The objective of this section is twofold: presenting the typical audiovisual processing steps in the case of video streaming, especially IPTV, and making the link between these processing steps and the overall integral quality.

The steps from extracting quality elements to deriving perceived quality are shown in Fig. 2.9. With the analysis of the processing steps, quality elements such as packet-loss can be identified. Typical degradations occurring at the different steps and the related quality features found in the literature will also be discussed in this section. Quality features are generally obtained by running analytical tests in which subjects

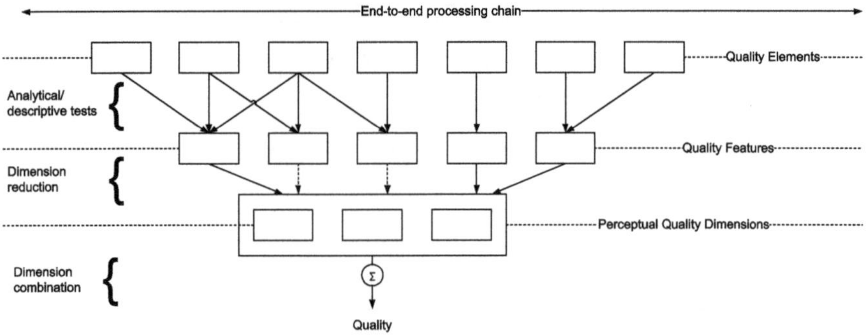

Fig. 2.9 From quality elements to quality

are asked to describe audio, video or audiovisual stimuli. Finally, the notion of quality dimensions, which define the sub-dimensional space that underly the quality features, will be introduced. These quality dimensions are generally obtained by conducting multidimensional scaling or semantic differential analysis techniques followed by a factor analysis.

Figure 2.9 also points out that several quality elements may contribute to one quality features and one quality element to several quality features. It also emphasizes that quality can be expressed as a linear combination of the quality dimensions. More details on this combination of quality dimensions is given in Sect. 2.2.3. With this structure which links quality elements to quality, and considering that quality elements can be measured or parameterized, quality can be predicted from measurements.

In the following, quality elements and features will be categorized by modality (audio/video/audiovisual) and by processing steps.

2.2.1 From Quality Elements to Quality Features

Figure 2.10 gives an overview of the processing steps applied to the audiovisual signal in the case of IPTV.

Processing steps are sequentially applied in three areas: the sender side (the IPTV content producer and provider), the transmission side, and the receiver side (the IPTV user). A coarse description of the processing steps is given in the following paragraph. Details are then provided per processing step in dedicated subsections. A general introduction to IPTV transmission can be found in Simpson (2008), and in a more concise way in Raake et al. (2011).

Audio and video signals are referred to as "source" signals when they have not yet been sent to the processing chain shown in Fig. 2.10. Typically, however, these signals have already been processed, at least during the recording phase. The source quality varies with the recording material. Source quality may also be reduced when

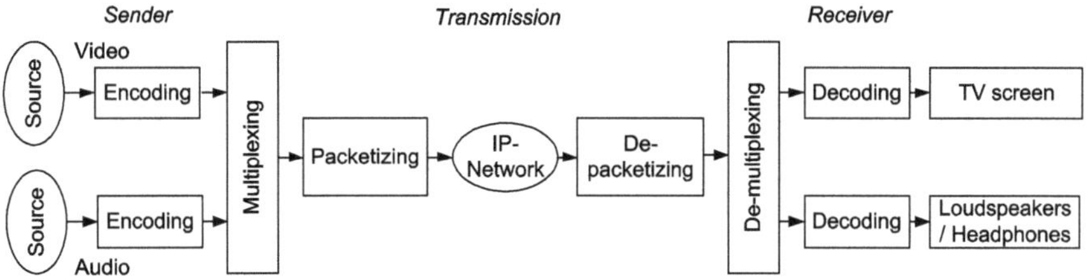

Fig. 2.10 Audiovisual processing steps for the considered application (IPTV)

transcoding is applied before transmission at the sender side. In the case of IPTV, this occurs for instance when the signal does not come directly from the producer, but is delivered to the TV Head-end via satellite. Specific to the audio signal is its sampling rate and the number of channels with which it has been recorded. The video signal is characterized by its resolution, its frame rate and its encoding format.

On the sender side, the audio and video signals are encoded separately. The resulting encoded signals are multiplexed, for instance in MPEG2-TS (ITU 2006), as commonly done in IPTV, before being encapsulated into packets, following a given packetization scheme. These packets are then sent over the Internet Protocol (IP) network. On the receiver side, packet handling mechanisms are applied for dealing with network delay and lost packets, then the inverse process is applied: de-encapsulation, de-multiplexing and decoding before playing out the audiovisual signal.

Video- and audio-specific quality elements, degradations and quality features are detailed below per processing step. They are also summarized per modality (audio, video, and audiovisual) and processing step in Tables 2.1 and 2.2. Note that the review of quality features actually lists both results from user tests and expert classifications of artifacts. Quality elements and features specific to the combination of audio and video into a multi-modal signal or stream are summarized in subsection "Audiovisual Quality Elements and Features".

2.2.1.1 Video Source

Describing the recording process is beyond the scope of this book. However, for quality assessment, it is of interest to study in which format the signal is usually recorded, and to analyse the potential sources of degradation. Detailed information can be found in Richardson (2002, Chap. 2).

The video signal consists in a succession of still frames, also called "images" or "pictures". Each frame is composed of a set of pixels and the number of pixels per frame determines the video resolution. Typical video resolutions for IPTV are PAL (720 × 576 pixels, used in Europe, and where 720 and 576 are the number of pixels per line and per column respectively), NTSC (720 × 480 pixels, used, e.g., in the United-States and in Asia), HD720 (1,280 × 720 pixels) and HD1080 (1,920 × 1,080 pixels). With SD and HD1080, typical frame-rates are 25 frames per second (fps)

Table 2.1 Quality Elements related with the different processing steps shown in Fig. 2.10

Source	Encoding	Multiplexing, packetizing	Transmission	De-multiplexing, De-packetizing	De-coding	Display
Video						
Picture size, frame rate, bit depth of sampling, chroma sub-sampling, scanning mode, pre-processing (frame-rate, frame size reduction), content characteristics (complexity, scene changes)	Codec type and profile, targeted bitrate, variable bitrate/constant bitrate, slice size, 8×8-transform usage, rate control, quantization, entropy coding	Multiplexing at transport stream-level/at real-time transport protocol-level, forward-error-correction	Packet-loss amount and pattern, jitter	Jitter-buffer, error correction, automatic repeat reQuest	Packet loss concealment	De-interlacing, color space conversion, up-scaling, display (size/aspect ratio, type (LCD, CRT), spatial resolution, etc.), viewing environment (viewing angle, illumination)
Audio						
Number of channels, sampling rate, content type	Codec type and profile, targeted bitrate, rate control, quantization, entropy coding	id. video	id.video	id.video	Packet loss concealment	Listening devices, rendering (mono/stereo), loudness, listening environment
Audiovisual						
Content type		Synchronization		Synchronization	Synchronization	

Details are provided in the text

Table 2.2 Quality features and artifacts from user tests and expert classifications of artifacts

Step	Features/artifact descriptors
Video capture	Smearing/blurriness, jerkiness (Yuen and Wu 1998), motion blur (Boev et al. 2009), noisiness.
Video encoding	Blockiness, blurriness, quantization noise, mosquito effect, (staircase effect, mosaic pattern effect, color bleeding, DCT bases-image effect) (Yuen and Wu 1998; Wolff and Ho 2006; Tek, ITU 1996a; Ries 2008)
Audio encoding	Quantization artifacts, pre-echo, aliasing artifacts, timbre distortion (birdies), binaural-unmasking distortion, muffled audio (band-limited artifacts), tone-trembling/sparking, tone-shift, noise overflow, tone spike, sawtooth, beat-artifact, tone leakage and blurred spatial position (loss of stereo image), clicking (Herre 1999; Erne 2001; Liu et al. 2008)
Video transmission	Jerkiness (Yuen and Wu 1998; ITU 1996a); slicing, freezing, blocking, ghosting (Greengrass et al. 2009)
Audio transmission	Interruptions, frame repetition (Perkins et al. 1998; Offir 2006).
Video display/reproduction	Combing effect (de-interlacing artifacts) (Le Callet et al. 2007); color space conversion artifacts; blurriness (Farias 2004; Yuen and Wu 1998; List et al. 2003); trade-off image size/compression artifacts (Le Callet et al. 2007; Jumisko-Pyykkö 2007); motion blur, lack of deep black, no depth feel (LCDs) (Tourancheau et al. 2007); color naturalness, sharpness, darkness, brightness, contrast, flicker, and smear/geometrical distortion (CRTs) (Teunissen and Westerink 1996)
Audio reproduction	Spatial-fidelity—(e.g., localization) and timbral-fidelity—(e.g., brightness) related-attributes.
Audiovisual transmission/decoding	Audio-video asynchrony, lip-synchronization.

Details are provided in the text

in Europe, and 29.97 fps in the United-States and Asiatic countries. A frame-rate of 24 fps is commonly used for cinema contents. With HD720, higher frame-rates are used (50 fps in Europe, 60 fps in the USA/Asia). Two scanning modes can be used to capture the video: *interlaced* or *progressive* format. In the case of progressive format, all lines of pixels are sampled per frame. In the case of interlaced format, odds- and even-lines are alternatively sampled by frame, and only half of the frame, also called “field”, is therefore captured.

Each pixel of the video frame is represented with three values indicating the proportion of Red (R), Green (G) and Blue (B) in the sampled pixel. The RGB signals are then transformed into another color space more suitable for coding, yielding one luminance signal component (Y) and two chrominance components (Cb and Cr), e.g., according to the standardized RGB to YCbCr color space transformation of the ITU-R BT-601 Recommendation (ITU 2011a). As the human visual system is less sensitive to color change than to luminance change, the two chrominance signals are then sub-sampled. For instance, with the YCbCr422 format, the encoder receives one luminance value for each pixel, and one chrominance value for every two pixels. With the YCbCr420 format, the encoder also receives one luminance value for each pixel, but one chrominance value for every 2×2 pixel-block. Note that YCbCr is often called YUV. Further note that each component (Y, Cb, Cr) is typically represented with 8 bits. More information on existing YCbCr/YUV formats can be found in http://www.fourcc.org/yuv.php.

Several types of degradations may appear at the capture level, or are linked to choices made during the recording process.

Directly linked to the "exposure time" of a camera, *smearing* of spatial detail occurs if there are moving objects in the direction of motion, resulting in a loss of spatial resolution and a blurring of features (Yuen and Wu 1998, Farias 2004). *Blurring* may of course also be caused by wrong focal setting or low-quality optics. *Motion blur*, at the source/capture-level, is generally caused by capturing in low light conditions.

The source material may also contain *noise*, due to the physical imperfections of the camera system. This kind of noise mostly becomes visible in dark scenes, where the amount of light hitting the camera sensors is very small.

Additional degradations may appear due to video-resolution- and frame-rate-conversion, or interlacing/de-interlacing processes. Indeed, the video frame-rate, resolution, scanning mode are impacting the number of pixels, and consequently the number of bits to be transmitted over the network. The color space conversion is one way of reducing the number of bits required for representing the video. If the same color space conversion is used for displaying the video, see paragraph "Video Post-processing and Display" later in this section, this process should not yield any artifact.

Using high video frame-rate and resolution ensures the *smoothness* and *sharp ness* of the video sequence. However, this also increases the demand in terms of bits to be transmitted. As a consequence, video frames may be skipped at the encoding stage in order to reduce the amount of video information that the system is required to transmit or process per unit of time. This may result in *jerkiness*, which is defined in ITU (1996a) as a "motion which was originally smooth and continuous, but is now perceived as a series of distinct "snapshots" of the original scene". It is often observed in the case of high motion scene.

This latter remark highlights that the scene characteristics may also indirectly impact the perceived video quality. For example, scenes with highly detailed structures or with complex movements require more bits for being encoded without introducing visible artifacts (see section "Video Encoding" for details).

Typical metrics used for characterizing the spatio-temporal complexity of the video contents are the Spatial perceptual Information (*SI*) and the Temporal perceptual Information (*TI*) measures of the ITU-T Recommendation P.910 (ITU 1999). *SI* is computed using Eq. (2.1):

$$SI = max_{time}\{std_{space}[Sobel(F_n)]\}. \tag{2.1}$$

Here, F_n is the luminance plane of the video frame at time n, and $Sobel(F_n)$ is its Sobel-filtered representation. std_{space} is the standard deviation over the pixels in the Sobel-filtered frame. max_{time} is the maximum value of the standard deviations std_{space} over the whole video sequence.

TI is computed using Eq. (2.2):

$$TI = max_{time}\{std_{space}[M_n(i,j)]\}. \tag{2.2}$$

where $M_n(i,j)$ is the difference between pixel values, i and j, of the luminance plane located at the same location in two consecutive frames. With the same naming conventions as for *SI*, $M_n(i,j)$ can be expressed as a function of time (n) as follows:

$$M_n(i,j) = F_n(i,j) - F_{n-1}(i,j) \tag{2.3}$$

Following the same computational steps as for *SI*, the standard deviation over space std_{space} of all $M_n(i,j)$ values is computed, and the maximum value over time max_{time} is taken as Temporal Information indicator. See ITU (1999) for details.

The number of scene changes in a video is also indirectly impacting the perceived quality of the video. Indeed, and as will be explained in section "Video Encoding", the video frame corresponding to the scene change is highly demanding in bits, which may temporally degrades the video. This degradation is however not perceivable if it does not last more than half a second (Yuen and Wu 1998). Moreover, and as will be explained in section "Video Encoding", scene changes may reduce the duration of the error propagation of loss when packet-loss hits a video frame.

In this work, the degradations occurring during the recording process and the color space conversion are assumed to be negligible compared to the degradations happening at later processing stage. It will also be considered that no pre-processing such as transcoding were applied to the signal.

2.2.1.2 Video Encoding

Despite the color space conversion, the bitrate required for transmitting uncompressed SD and HD video signals is too high for being transmitted without compression. In IPTV, MPEG-2 Part 2 (also referred to as "H.262") (ISO/IEC 2000; ITU 2012a) and H.264 (also referred to as "MPEG-4 Part 10" or "MPEG-4 AVC" (Audio Advance Coding) (ISO/IEC 2010; ITU 2012b) codecs are typically used for encoding the video signal. In current services, H.264 is mainly used, due to its higher performance. MPEG-2 is only used for reasons of backward compatibility. Both codecs follow the same encoding principles, which are described in details in Richardson (2002) and Wu and Rao (2006b).

During the encoding process, video frames are assigned different types, which are called "I-frames", "P-frames" and "B-frames". "B-frames" are further categorized between non-reference B-frames and reference B-frames, the latter being used in the case of hierarchical coding. "I" refers to intra-coded frames, i.e., an I-frame does not require information from other frames for being coded, while P- and B-frames are predicted in a temporal manner from neighboring I-, P- frames or reference B-frames

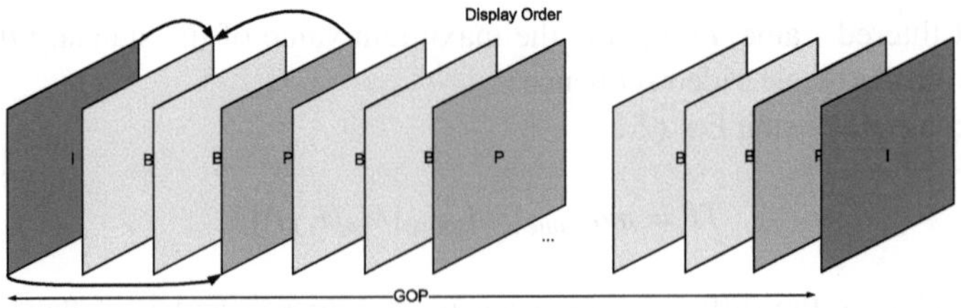

Fig. 2.11 Video encoding process

(see Fig. 2.11). P-frames are unidirectional: they are predicted only from previous frames, while B-frames are bidirectional and are therefore predicted from previous and/or subsequent I-, P-, or reference B-frames. The set of frames beginning from an I-frame up to the next I-frame (excluded) is referred to as a Group of Picture (GOP) (see Fig. 2.11). There is commonly one I-frame per second or half a second in IPTV services.

When the GOP is "open", frames belonging to one GOP may be encoded using reference frames from the following GOP. When the GOP is "closed", only reference frames from the current GOP can be used as reference.

The encoder receives the YUV420 (also called YCbCr420) signal. This signal is then processed in blocks of 16×16 pixels area, called macroblocks, which are represented by 16×16 "Y" (luminance) values, 8×8 "Cb" values and 8×8 "Cr" values. Different types of transforms can be applied on the macroblocks. An MPEG-2 encoder uses the 8×8 Discrete Cosine Transform (DCT): four DCTs on the luminance signal and one DCT on each of the chrominance signals. An H.264 encoder usually applies sixteen 4×4 integer transforms on the luminance signal, and four 4×4 integer transforms on the two chrominance signals. This 4×4 integer transform has only coefficients of 0, 1 and 2 and can therefore be implemented without multiplication. Still, its performance is almost identical to the performance of the DCT. For intra coding, H.264 has also an 8×8 integer transform available. All transforms follow the same principle, as described below for an 8×8 transform.

Each 8×8 block of pixels is transformed into a block of 8×8 DCT coefficients which describe the spectral behavior of the block, where sharp edges and sudden spatial changes represent high frequencies. In areas with few to average spatial complexity, i.e., few details, most coefficients are therefore null, reducing the amount of information to be transmitted.

In the following step, the coefficients are quantized, the quantization step depending on the available transmission bitrate. This is a crucial step, since the loss of information due to quantization cannot be recovered. As a consequence, if the quantization is too coarse, the coefficients are either fluctuating too much from their original value, or are even truncated to zero, yielding several artifacts, as will be described below.

In the case of P- and B-frames, the blocks are not directly transformed as for I-frames. Instead, a prediction of the macroblock to be coded is generated, and only the prediction error, i.e., the difference between macroblock and prediction, is further processed (i.e. integer transformed in the case of H.264). This prediction error is obtained by searching for the most similar area of pixels in one of the previous or subsequent (B-frame only) I- and/or P-frames. The displacement between the position of the current macroblock and its most similar equivalent in the chosen reference frame is described by a so-called "motion vectors", which also need to be transmitted in the bitstream for reconstruction of an identical signal at the decoder side. In MPEG-2, motion vectors always point to 16×16 areas, whereas in H.264, motion vectors can also address areas of down to 4×4 pixels.

In the case of H.264, macroblocks of one video frames are arranged into slices. Macroblocks within one slice are coded in raster order. Three main types of slices handling are used: "I-", "P-" and "B-slices". I-slices contain only I-macroblocks while P- (resp. B-) slices contain P- (resp. B-) and I-macroblocks. I-macroblocks are predicted using intra prediction from decoded macroblocks or 4×4 blocks in the current slice. P- and B-macroblocks are predicted from reference frames.

Note that each slice does not require information from the other slices of the picture for being decoded. The slice typically contains a fixed number of macroblocks, for instance one macroblock-row. In this case, the slice size in (encoded) bits may vary depending on the picture content (i.e., spatio-temporal complexity). Alternatively, slices can be determined in terms of fixed amounts of encoded bits, corresponding to varying spatial areas of a given frame (again, depending on the spatio-temporal complexity of the different areas).

For further reducing the amount of data to be transmitted, the resulting quantized coefficients are then entropy-encoded using Run-Length Coding (RLC), Variable-Length Coding (VLC) or even Context-Adaptive Binary Arithmetic Coding (CABAC) as in H.264. This entropy coding is lossless. Among these coding tools, CABAC is the only one to be dependent of the codec profile. It is indeed not available in the H.264 baseline profile.

An important demand on video compression for many applications is to deliver a Constant BitRate (CBR). For average videos with a largely fluctuating spatio-temporal complexity, this demand can only be fulfilled by varying the amount of compression and therefore the output quality. As a result, highly complex parts of the video will have lower visual quality than easy-to-code parts.

The only method to systematically vary the compression rate in an encoder is changing the quantization step size. In the case of CBR, the quantization step size is only fixed per macroblock, but can vary within a single video frame or even through the whole coded sequence. It is selected as a function of the required output bitrate of the encoder (the "target" bitrate), the frame type and the spatio-temporal complexity of the current video scene. A buffer is used for absorbing remaining bitrate variations. The buffer fullness state is used as feedback information for future quantization step adjustments. This whole process is called "rate control".

Without any rate control, i.e., with constant quantization step size, the data rate of the encoded signal would be barely unpredictable, but the visual quality would remain almost constant. This mode is called Variable BitRate Coding (VBR) and may be used for encoding video on DVDs.

More information on the encoding process can be found in Richardson (2002). For more details on the H.264 codec, the reader is also referred to the ITU-T H.264/MPEG-4 Part 10 standard (ITU 2012b; ISO 2010), and the corresponding over-view in Wiegand (2003).

Substantial work has been conducted for identifying degradations linked to video encoding. An overview of these degradations and related literature is listed in the row "video encoding" of Table 2.2. In the following, most commonly cited video encoding degradations will be detailed. For further information on these degradations, the interested reader is referred to the literature cited in Table 2.2, and especially to Yuen and Wu (1998).

Blockiness, also referred to as *block distortion* or *tiling* (ITU 1996a) is a distortion of the image characterized by the appearance of an underlying block encoding structure. Block distortions are caused by coarse quantization, and the "blocky" appearance reflects the underlying block structure of MPEG video coded images. Block distortions comprise other identified degradations such as the *staircase effect*, *mosaic pattern effect* or the *DCT basis-image effect* (Yuen and Wu 1998). In modern encoders, coarsely quantized block boundaries are usually filtered, reducing the visibility of the above artifacts but yielding blurriness (see below).

Blurriness is a more global distortion of the entire image or image areas, characterized by reduced sharpness of edges and spatial details (ITU 1996a). It commonly occurs with the use of low bitrate algorithms, and/or in higher activity regions of images, such as in areas with highly detailed textures or around object edges and/or during high-motion video (Telchemy Incorporated 2008). It is caused by suppression of high-order DCT coefficients through coarse quantization or block-based filtering to reduce blockiness (List et al. 2003).

Another type of artifact caused by strong quantization of high frequency coefficients, but this time chrominance coefficients, is *color bleeding*. Since chrominance is typically sub-sampled, bleeding can occur beyond the range of a block.

Mosquito noise is a temporal degradation resulting from the different encodings of the same area of a scene in consecutive frames. It is characterized by moving artifacts and/or blotchy noise patterns superimposed over the objects. This artifact is the most visible over moving edges. It resembles a mosquito flying around a person's head and shoulders (ITU 1996a).

Finally, *quantization noise* is a "snow" or "salt and pepper" effect similar to a random noise process but not uniform over the image (ITU 1996a). It is created in the quantization step during the encoding process and it is caused by the coarseness of the quantization levels.

2.2.1.3 Audio Encoding

Audio codecs addressed in this work are MPEG-1 Layer 2 (MP2) (ISO/IEC 1993), MPEG-1 Layer 3 (MP3) (ISO/IEC 1993, 1995; Brandenburg et al. 1992), Advanced Audio Coding (AAC), which is part of both MPEG-2 and MPEG-4 specifications (ISO/IEC 2006a, b; Bosi et al. 1996), MPEG-4 High Efficiency Advanced Audio

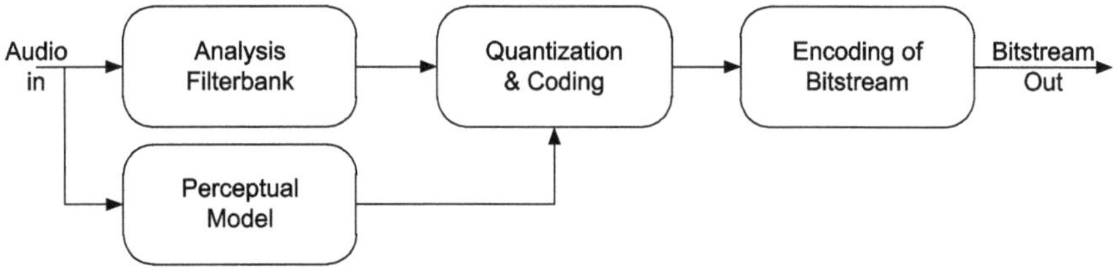

Fig. 2.12 Generic structure of a perceptual audio encoder, reproduced from Herre (1999) and Brandenburg (1999)

Coding (HE-AAC) (ISO/IEC 2006b) and Dolby AC-3 (AC-3) (Advanced Television Systems Committee 2010; Davis 1993). MP2 is used in this work as baseline codec. Most commonly used codecs in audiovisual streaming are AAC, HE-AAC and AC3. In the ITU-T P.NAMS and P.NBAMS competitions, all these codecs were used in the tests, apart from MP3.

Most of the audio codecs take advantage of the following facts: not all frequencies are simultaneously present in a typical audio signal, not all sounds are perceivable, and some sounds are masked by other sounds, including both spectral and temporal effects [see Painter and Spanias (2000) for more information on these psychoacoustics principles].

MP2, MP3, AAC, HE-AAC and AC3 codecs belong to the filter-bank-based codec family and follow similar encoding steps, also shown in Fig. 2.12:

- "Analysis Filterbank": Bandpass filterbank processing, which divides the signal spectrum of the signal into N [e.g., N = 32 ISO/IEC (1993)] frequency sub-bands for each measurement window. This results in a set of coefficients which represent the frequency localized signal power within each sub-band (Painter and Spanias 2000).
- "Perceptual Model": Psychoacoustic modeling of the audio signal in spectral bands. The time- and frequency-dependent masking threshold is estimated per sub-band. In other words, the perceptual model outputs values for the masking threshold, and therefore the allowed quantization noise for each sub-band.
- "Quantization and Coding": Bit allocation for each sub-band sample and based on the output of the perceptual model, with the target of an optimal use of available bits in terms of the per-band quantization noise, considering the masked threshold of that noise per band.
- "Encoding of bitstream": the encoded signal (coded spectral values and additional side information) is entropy coded, then packed into the bitstream to be transmitted.

The PCM audio signal is first filtered into N bands.

In the case of MP2 and MP3, a polyphase filter bank is used, transforming a block of 1,152 samples into 32 equal-width frequency sub-bands (Pan 1995). In the case of MP3 (Brandenburg et al. 1992; Brandenburg 1999), a Modified Discrete Cosine Transform (MDCT) is applied to each sub-band, resulting in a more accurate frequency analysis of the signal (see Fig 2.13). Each of the windows used during

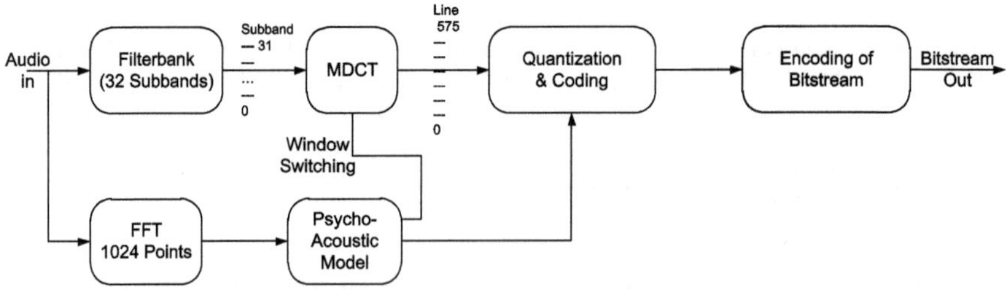

Fig. 2.13 Simplified diagrom of MP3, adapted from Brandenburg (1999)

the MDCT are overlapping two consecutive sub-bands. The type of window used during the MDCT is determined by the psychoacoustic model. The MDCTs result in 18 coefficients per sub-band, therefore 576 MDCT coefficients for the whole audio frame. Sub-bands are here corresponding to the critical bands of the human hearing.

Based on a 1,024-point Fast Fourier Transform (FFT), the psychoacoustic modeling block outputs masking thresholds per sub-band. This threshold is compared to the maximum signal level in the sub-band for computing the Signal to Masker Ratio (SMR) for this sub-band. The SMR values are used for determining the number of bits to be allocated per sub-band. The maximum absolute value of the samples in the sub-band is quantized to 6 bits. This is the *scale factor* for the sub-band. The sub-band samples are linearly quantized to the bit allocation for that sub-band.

The basic coding paradigm of MPEG-2/4 AAC is the same as MP3's (Brandenburg 1999), but it improves MP3 in many details. In particular, it uses new coding tools for improving the quality at low bitrates (Brandenburg 1999). In addition, AAC uses the MDCT filterbank directly. There was indeed no need for combining polyphase filter bank to MDCT filter bank, apart from compatibility with the MP2 structure.

In the case of MP3 and AAC, the MDCT coefficients are Huffman-encoded before being packetized and transmitted.

The last step before transmission consists in packetizing the bit allocation table, the scale factors and the sub-band (MP2) or MDCT (MP3, AAC) samples into the bitstream.

HE-AAC (ISO/IEC 2006b; Meltzer and Moser 2006) combines the MPEG2/4 AAC technology to the spectral band replication (SBR) technology, which enhances the compression efficiency in the frequency domain, and to the Parametric Stereo tool (HE-AACv2), which increases the compression efficiency of stereo signals. Indeed, AAC is yielding transparent quality at typical bitrate of 128 kbps (Meltzer and Moser 2006). However, at lower bitrate, the quality would decrease. The HE-AAC codec has been designed for being used at bitrate lower than 128 kbps, typically 64 kbps down to 24 kbps.

In HE-AAC, the AAC codec is responsible for coding and transmitting the lower and mid-frequencies of the spectrum only. The SBR tool generates the higher frequencies. SBR does not transmit the full spectrum but reconstructs the higher frequencies in the decoder based on an analysis of the lower frequencies transmitted in AAC, and on the envelope of the spectrum, which is transmitted in a parametric manner. The reconstruction of these high frequencies is optimized for harmonic and noise-like components.

With HE-AACv2 at low bitrates, only one channel is transmitted in the case of a stereo signal. When Parametric Stereo is used in HE-AAC, side information is transmitted along the encoded mono channel in order to describe the spatial intensity stereo generation and ambience regeneration at the decoder. This technique performs well only at low bitrate, i.e., below 32 kbps.

More details on HE-AAC can be found in Meltzer and Moser (2006).

With the AC-3 codec (Davis 1993; Advanced Television Systems Committee 2010), the PCM audio signal is divided in blocks of 512 samples. Blocks are overlapping, so that the 256 last samples of a block are also the 256 first samples of the following block. Each block is transformed from the time-domain to the frequency domain in the analysis filter bank. This transform results in 512 frequency coefficients. Each coefficient is represented as a binary exponent (the quantized values) and a mantissa (the scale factor). The set of exponents is further encoded into a coarse representation of the signal spectrum which is referred to as the spectral envelope. The spectral envelope is used for determining how many bits shall be used per mantissa. The encoded envelope and the quantized mantissa are then transmitted. Six consecutive blocks are analysed in a row, forming one AC-3-frame.

Typical audio artifacts resulting from the encoding process are quantization distortions, pre-echo artifacts, aliasing artifacts, timbre distortions, binaural unmasking distortion, and loss of stereo image (Erne 2011; Liu et al. 2008). Detailed descriptions of the main artifacts are given in the following.

Pre-echo (Herre 1999; Erne 2011; Liu et al. 2008) may occur with time-domain transient signals when long block size windows are used for transform coding. Long block sizes are preferred for coding efficiency purposes, since less side-information is then required, and slow-varying signals can be coded more efficiently. However, in the case of time-domain transient signals such as castanets, cymbals, drums, or fricative and plosive in the case of speech, quantization noise spreads over the whole analysis window. Indeed, time-domain transient signals spread out over many subbands in the frequency domain, and quantization noise could be masked during and after the transient due to spectral and temporal (post-) masking. During the decoding process, the quantization noise—which was supposed to be masked—spreads over the complete block, resulting in an audible "pre-echo" of the upcoming transient event, or in speech reverberation artifacts in the case of fricatives and plosives. This artifact can be reduced by using adaptive block-switching, and therefore adapting the size of the block to the analysed signal, or by Temporal Noise Shaping (TNS), as described in Herre (1999).

Aliasing artifacts (Erne 2011; Liu et al. 2008] are inherent to the use of filter-banks. As a consequence, aliasing cancelation such as time-domain aliasing cancelation (TDAC) are generally applied, as for instance in MP3, AAC and AC-3.

As previously mentioned, the bit allocation varies from one block (of, e.g., 512 samples) to another block. As a consequence, spectral coefficients may temporarily appear and disappear, resulting in a change in timbre and in a variation of high frequency energy. Due to compression constraints, high-frequency content may be cut yielding *muffled* audio (Liu et al. 2008). These two above phenomenons are referred to as *birdies* and *band-limitation* artifacts (Liu et al. 2008; Erne 2011).

The masked threshold of a signal can sometimes be lower when listening with two ears rather than one. In particular, the detection of a signal in noise improves when either the phase or level differences of the signal at the two ears are not the same as the masker. This phenomenon is called the *binaural masking level difference* (BMLD), or *binaural unmasking*. An implication of this is that the signal and masker appear to originate from different locations in space, and that quantization noise thus may be audible.

In the case of HE-AAC(v2) where spectral band replication (SBR) and Parametric Stereo (PS) are applied, additional artifacts may occur (Liu et al. 2008). With SBR, high-frequency bands are derived from replicated low-frequency bands. Changes in the replicated source are thus resulting in spectral discontinuity in the sub-bands. This is perceived as (tone-)*trembling* or *sparkling*. The direct replication of low bands may also result in a *tone-shift*, but this phenomenon is usually not perceivable. More critical, especially for tone-rich signals, is the *noise overflow* and the *tone spike* artifacts, which are perceived as rasping and metallic sounds. They result from inaccurate tonality matching. The limiter gain mechanism in SBR prevents results in the unwanted restriction of some scaling gains in the high-frequency bands. This is the *sawtooth* phenomenon. Lastly, the patch procedure of the SBR may result in the *beat*-artifact, which consists in an amplitude fluctuation at a regular rate, resulting from the mutual interference of two close tones.

As previously mentioned, the Parametric Stereo (PS) module reconstructs a stereo signal from the monaural down-mixed signal according to the parameters extracted by capturing the stereo image of the input binaural signal. This down-mixing procedure may result in the *loss of stereo image*, while the *tone leakage* artifacts, which refer to the leaking or vanishing of one channel to another channel, originate from the variability of mixing coefficients. These phenomenons are perceived as *blurred spatial position*. Additional artifacts such as *clicking* are also linked to the PS module. More information can be found in Liu et al. (2008).

2.2.1.4 Multiplexing and Packetization

The audio and video encoding process has now been described. This results in encoded audio and video streams referred to as Elementary Stream (ES). Next processing step consists in packetizing the elementary streams into Packetized Elementary Stream packets (PES packets). Each Packetized Elementary Stream (PES) packet starts with a PES header, and each PES packet header contains timing information relevant at the receiver side for playing out the data. In particular, time stamps in each PES can be used to ensure lip-sync between the video and audio, to order frames at the receiver side into the correct display order, and to identify the exact number of lost frames (see also section "De-packetization and De-multiplexing").

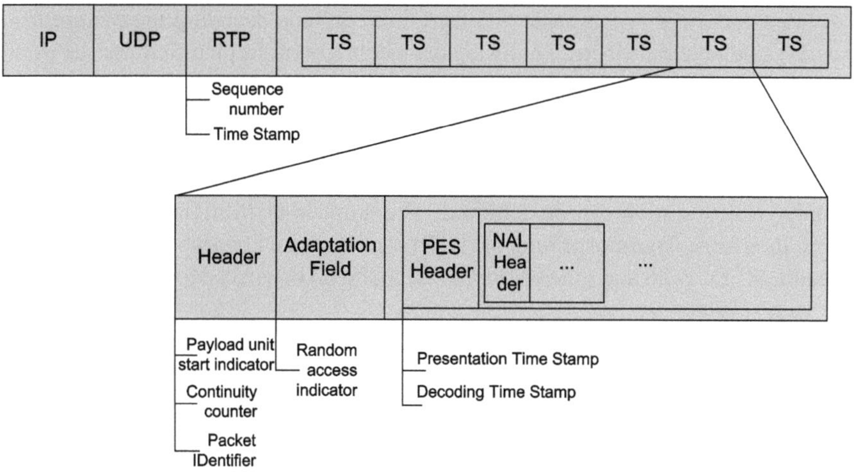

Fig. 2.14 The TS/RTP/UPD/IP packetization scheme with H.264 video Elementary Streams

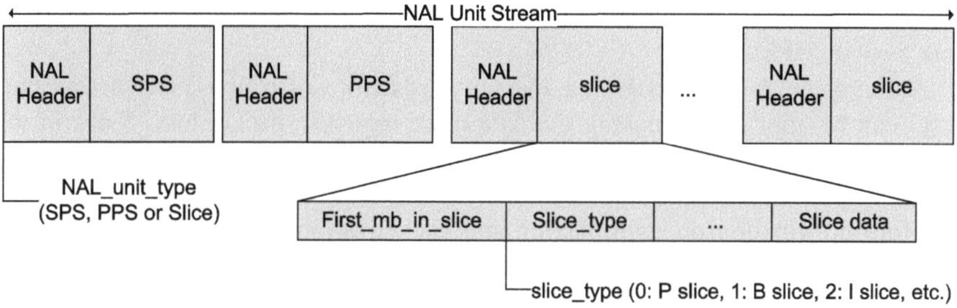

Fig. 2.15 NAL unit stream. See text for details

The TS/RTP/UPD/IP packetization scheme for H.264 is drawn in Fig. 2.14. Details are provided in the following. Packet headers are explained in section "De-packetization and De-multiplexing".

In the case of H.264, the ES is packetized into Network Adaptation Layer (NAL)-units (see Fig. 2.15). The NAL-unit streams contain several NAL packets. The two first NAL packets, which are required for decoding the video, contain information on the current video sequence (SPS: Sequence Parameter Set) and the current pictures (PPS: Picture Parameter Set). The remaining NAL packets mainly contain video slices. A complete video slice is always contained in one or more complete NAL units. As previously mentioned, each slice does not require information from the other slices of the picture for being decoded. It can be parsed separately from the bitstream, which is achieved by adding a respective slice header.

The use of NAL units makes H.264 more robust to packet loss. Indeed, since each NAL unit contains one or more complete slices, it also carries a re-synchronization point of the H.264 decoder. As a consequence, if an IP-packet is lost, the decoder can immediately re-synchronize and therefore continue decoding the picture from the next available slice instead of losing the whole rest of the picture frame, as would be the case without the use of slices.

In IPTV, the PES is often packetized further into MPEG2-TS packets. Each MPEG2-TS packet contains audio-only, video-only, or other stream-related data.

With the multiplexing of audio and video TS packets into one RTP packet, audio-visual synchronization can be achieved. The number of MPEG2-TS packets in a single Real-time Transport Protocol (RTP) packet is usually seven. The proportion of audio MPEG2-TS and video MPEG2-TS in the RTP packet depends on the audio and video bitrates.

Finally, each RTP packet is encapstulated into User Datagram Protocol (UDP) and IP. Each packetization step adds packet headers to the payload. The size of the resulting IP packet, which contains seven TS-packets, is slightly less than the Standard Ethernet-Maximum Transmission Unit (MTU) size (1,500 Bytes).

Other packetization schemes than the just described TS/RTP/UPD/IP packetization scheme (Fig. 2.14) such as RTP/UDP/IP, or TS/UDP/IP may also be used in IPTV. In the absence of MPEG2-TS, the multiplexing is done at the RTP level. In that case, one RTP packet contains only one type of data, audio or video, and not a mix as with MPEG2-TS.

At this stage, further processing, like Forward Error Correction (FEC) at transport level can be applied for making the data more robust to packet loss. Since in the following only the resulting packet loss at the player is of interest, this mechanism will not be described in this work.

More information on multiplexing and packetization in the case of IPTV can be found in Simpson (2008), ITU (2006). A detailed description of NAL units is provided in Wiegand (2003).

2.2.1.5 Transmission

Once encapsulated, IP packets are transmitted over the IP-network. During transmission, packet loss and jitter can occur. Packet loss is usually caused by network congestion due to overload in the network. Packets of the same stream may take different paths through the network. This may cause highly diverging delays, referred to as packet jitter. Indeed, even though packets are re-ordered, packet jitter needs to be compensated by the use of a sufficiently large input packet buffer. If this buffer runs empty, there is temporally no information to be decoded, and *jerkiness* occurs (see section “Video Source” for a definition of “jerkiness”).

The residual packet loss at the player can be characterized by the packet-loss rate and pattern. Since losing several packets in a row may have a different quality impact than losing several times a single packet, information about the packet loss pattern is useful for describing the impact of packet loss on the perceived quality.

Popular models for representing loss pattern observed from real-traffic are the 2- and 4-state Markov Models (MMs) (Sanneck and Carle 2000; Bolot et al. 1999), as well as the 2-state Gilbert-Elliot model (Gilbert 1960; Elliott 1960). These models can be used for realistically impairing encoded audiovisual streams. Dropping packets according to such state models allows a straightforward manipulation of the gap and burst period, where a gap is defined as a period with high density of received packets, while the burst represents a period with high loss density. With the 4-state Markov Model (MM)-model and the 2-state Gilbert-Elliot model, probabilities of found packets in a burst period, and vice versa, can be tuned.

2.2.1.6 De-packetization and De-multiplexing

At the receiver side, the IP stream is de-packetized and de-multiplexed by parsing packet headers. This results in audio and video ESs to then be decoded. In the case of the TS/RTP/UDP/IP packetization scheme, the packet header parsing is entirely described in the ITU-T P.1201.2 standard (2012e). This standard has been developed for PES- and TS-encrypted bistream. Additional information may also be found in the IETF RFC 2250 (1998), IETF RFC 3550 (2003), and ITU-T H.222 (2006) documentations.

Packet header information is acquired from the RTP, TS, and PES headers, if they are not encrypted (see Fig. 2.14). They are then used for extracting information on audio and video frame boundaries and losses, and on video frame types.

For instance, the Sequence Number of the RTP header is used for re-ordering the RTP packets. It can also be used for detecting lost packets, and therefore computing the RTP-packet-loss percentage or the averaged number of consecutively lost RTP packets, referred to as "burstiness" at RTP level. The RTP headers also contain time stamps which change for each PES frame and can therefore be used for detecting lost PES frame boundaries [see ITU (2012e) for details]. Changes in the Decoding Time Stamp (DTS) of the PES header can also be used to detect lost PES frame boundaries, but the DTS time stamp has lower accuracy than the RTP time stamp.

At the TS header level, the Packet IDentifier (PID) indicates if the TS payload contains audio or video, and the continuity counter (CC) keeps track of the TS packets over different RTP packets. Combined with timing information from the Presentation Time Stamp (PTS), the PID and the CC of the TS headers can be used for computing the percentage of audio or video lost TS packets and the average number of consecutively lost audio or video TS packets. The audio and video bitrates can also be derived from this information.

The *payload_unit_start_indicator* of the TS header indicates audio and video frame boundaries. Combined with the payload size, this indicator can be used for computing audio and video frame sizes.

The *random_access_indicator* of the TS adaptation field indicates the start of an I- or Instantaneous Decoder Refresh (IDR)- frame.[2] This information is especially useful for the detection of GOP-boundaries and computing parameters such as the

[2] An IDR frame is a specific type of I-frame in H.264. If the decoder receives an IDR frame, it can refresh its buffer since no frame after the IDR frame can reference any frame before it.

video loss duration or the spatio-temporal complexity of the content (see Sect. 5.3 for details).

2.2.1.7 Packet-Loss Recovery Techniques

Packet-loss recovery techniques can be classified into two groups: the sender-based and the receiver-based techniques. Both techniques are complementary. Sender-based methods such as FEC and retransmission are targeting lowering the overall percentage of packet loss by correcting lost packets. When the sender-based methods have failed to correct all lost packets, receiver-based methods—or error concealment—may be used. In that case, the percentage of packet-loss stays unchanged after the error concealment is applied, but the impact of the perceived degradation is reduced by decreasing the error due to the missing content. The type of the substitute content depends on the error concealment technique.

Sender-based methods are beyond the scope of this book and are not further described in this work. The interested reader will find more information on this topic in Perkins (1998) and Korhonen (2006).

The receiver-based methods are described in the following "Video Decoding" and "Audio Decoding" sections.

2.2.1.8 Video Decoding

At the decoder stage, the video is decoded by reversing the process applied during encoding, i.e., by applying bitstream parsing, reversed entropy coding and inverse transforms (DCT or integer).

In the case of packet loss, the corresponding perceptual impact depends on the type of the frame in which the loss occurs and on the applied packet loss concealment technique. Indeed, P- and B-frames are predicted from previous I- and P-frames while I-frames are intra-coded and thus do not depend on previous frames. As a consequence, if a loss occurs on an I- or a P-frame, the loss is typically propagated till the next I-frame. If a loss occurs on a reference B-frame, which is used in hierarchical coding, the loss propagates till the next P- or I-frame, i.e., it only affects the surrounding non-reference B-frames. There is no loss propagation if the loss occurs on a non-reference B-frame.

Moreover, if *slicing* is applied, one packet loss results in the loss of the corresponding pixel-area as well as the pixel-area corresponding to the rest of the last affected slice. The decoder re-synchronizes its bitstream parsing process at the beginning of the next slice, using the slice header, and the rest of the frame can correctly be decoded, if no additional loss occurs in the frame. Note that the spatial extent of the loss depends on the number of slices per frame. This point will be re-addressed later.

Blocking effect may originate from packet loss in case that slicing is applied. Indeed, when loss occurs, content from the previous reference frame is usually copied to the lost portion of the hit frame. If there is motion in the sequence, this

Fig. 2.16 Blocking artifacts resulting from packet loss and in case that slicing is applied. Source content (without blocking artifacts) comes from the Blender open movie project "Sintel"

replaced content will not fit well the missing content, and the whole lost block of pixels will be displaced, resulting in a blocking artifact. See Fig. 2.16 for an example of *blocking* artifact in the case of packet loss, and when slicing is applied.

Another loss handling strategy is *freezing*. In this case, the frames affected by losses and the frames referencing the lossy frames are completely discarded and replaced by the last unimpaired reference frame till the next unimpaired I-frame.

Finally, another type of degradation resulting from packet loss is *ghosting*. Ghosting occurs when packet loss removes an I-frame, or a large chunk of slices within an I-frame, which is placed at a scene change. In that case, a picture before the scene change may be used as concealment for that I-frame. As the decoder receives the P- and B-frames from the new GOP (the new scene), it will add this new information to the lossy I-frame, falsely containing the old scene, and thereby creating the ghosting effect in the frames of the new GOP (Greengrass et al. 2009).

2.2.1.9 Audio Decoding

The Audio Decoding process is summarized below for the different codecs.

In the case of MP2, the following steps are applied: (a) scale factor decoding, (b) inverse quantization, using the bit allocation data, (c) inverse compression of the sub-band samples, and (d) filter-bank synthesis. In the case of MP3 and AAC, MDCT values are Huffman decoded at step a) before being inverse quantized.

In the case that packet-loss-recovery has failed to recover lost packets (if any), error concealment techniques are applied. As described in Perkins (1998), there are three types of error concealment methods: the insertion-based methods, the interpolation-based methods, and the regeneration-based methods—or compressed-domain-interpolation-based method.

In the case of IPTV and MPEG2-TS-based transmission, the audio is multiplexed with video information into one IP-packet. Considering the relatively low bitrate of audio, compared to the video bitrate, the number of audio TS-packets within one IP-packet can be considered as low for all realistic cases. As a consequence, a given audio frame is distributed across one or typically multiple packets. Hence, in the case of bursty packet loss, the number of subsequent frames affected by the loss is typically small. Thus, in actual IPTV implementations, no complex error concealment methods are employed; these are typically of the insertion-type. For information on the other types of error concealment, see Offir (2006), Perkins et al. (1998).

With the insertion-based method, the audio content corresponding to the lost packet(s) is replaced by silence (*interruption*), noise, or the previous found packet(s) (*frame repetition*). In another version of this method, the missing content is not replaced at all, the missing frames are skipped. This results in a shorter audio stream, and therefore *asynchrony* in the case of an audiovisual signal. All variants of the insertion-based methods are easy to implement but their effectiveness decreases with the increase of loss duration.

2.2.1.10 Video Post-processing and Display

If the video was interlaced, de-interlacing is applied on the decoded video stream.

The process of interlacing causes a well known artifact which manifests itself as a series of horizontal lines and is often referred to as the *combing effect*, as the lines look like the teeth of a hair comb. By design, interlaced video is at its best quality if it is captured, transmitted, and displayed in the same interlaced format. Interlaced displays such as CRTs (Cathode Ray Tubes) have been replaced by progressive displays such Liquid Crystal Display (LCD) displays. However, many media providers still produce or transmit interlaced video. As a consequence, interlaced video need to be de-interlaced, which results in a loss of quality. Since each frame of interlaced video is composed of two fields that are captured at different moments in time, and each field either carries the even or the odd lines of the video frame, interlaced video frames will exhibit the combing effect if the recorded objects are moving fast enough to be in different positions when each individual field is captured. On former CRT-displays, the fade out time was large enough, and the actual resolution was low enough so that the combing effect was not visible. For modern display technologies, even a good de-interlacing engine, which is necessary in order to drive modern progressive scan displays such as LCDs, is hardly able to completely remove the combing effect.

Another processing step is to convert the YCbCr values back to RGB values. This step is an (almost) lossless matrix conversion and usually does not have a visible influence on the perceived quality. Depending on the processing chain, this step may be done by the monitor itself. Note that if the RGB-converter does not use the same color space conversion as the one used at the sender side, color artifacts may appear.

In the case of SD (Standard Definition, PAL or NTSC), the video is also often up-scaled to HD (High Definition, HD720 or HD1080), since most consumers have

HD screens, even though up-scaling may result in additional artifacts. At least, the SD-video will appear visibly *blurred* on the mostly large HD-screen. It is also interesting to see the trade-off between video resolution (*image size*) and degradations (*compression*), as reported in Le Callet et al. (2007): for low degradations, the High Definition (HD) resolution is preferred to Standard Definition (SD). This preference inverses with increasing degradations.

At the time they were introduced to the market, LCD displays were not outperforming the CRT screens in terms of quality, since they were criticized for the appearance of *motion blur* for fast-movement video contents, the *lack of deep black*—due to too high luminance values at black level—and *no depth feel* (Tourancheau 2007). Nowadays, LCD screens are dominating the 2D-TV-screen market. Even though CRTs are no longer used, it is worth mentioning that Teunissen et al. (1996) found in their study that, for CRTs, eight quality features are relevant for the application: *color naturalness*, *sharpness*, *darkness*, *brightness*, *contrast*, *flicker*, and *smear/geometrical distortions*. More details will be provided in Sect. 2.2.2.

2.2.1.11 Audio Reproduction

In the context of IPTV, audio is usually played out via loudspeakers. With the use of headphones, the influence of the room may be considered as negligible (EBU 1998), but users are more sensitive to audio artifacts (Schobben and van de Par 2004). Different types of headphones exist: closed-back *circum-aural* headphones give the user the impression that the sound is coming from within his/her head. This effect appears with open-back headphones as well, and these latter present lower noise-reduction. *Circum-aural* headphones, which encompass the outer part of the ear (the pinna), attenuate outside noise more than *supra-aural* headphones, which are placed on the top of the ears, but yield more discomfort. More information on the types of headphones and respective tests can be found in Bech and Zacharov (2006), which refers to the ITU-T Rec P.57 and IEC 60268-7 Recommendations.

If loudspeakers are used, several additional quality elements – related to the room environment—have to be considered: the room dimensions and shape, the reverberation time, the background noise level, the attenuation of early reflection, and the suppression of flutter echoes.

The number of reproduction channels and respective spatial reproduction technique applied, the listening level, the position of the loudspeakers and of the listener(s) are also impacting the perception of the audio signal.

Perceptual features linked to the audio reproduction system can be categorised into spatial fidelity and timbral fidelity related attributes (Rumsey et a. 2006). Examples of spatial fidelity attributes are "spaciousness" (Zacharov and Koivuniemi 2001; Choisel and Wickelmaier 2007), which refers to the perceived size of the environment, or "localization" (Guastavino and Katz 2004), i.e., the perceived sound location. The perceived "volume" (Zacharov and Koivuniemi 2001) and "brightness" (Choisel and Wickelmaier 2007) are examples of timbral-fidelity attributes. The interested reader is referred to Zacharov and Koivuniemi (2001), Guastavino and Katz (2004),

Choisel and Wickelmaier (2007) for more details on spatial fidelity and timbral fidelity attributes.

2.2.1.12 Audiovisual Quality Elements and Features

All audio and video quality elements and features can be considered in terms of their audiovisual interaction and contribution to audiovisual perception. Another important and well-known audiovisual quality feature is the *synchronization* between audio and video quality (Jumisko-Pyykkö 2007; Hollier et al. 1999; ITU 2003b). Synchronization, and especially *lip-synchronization* is usually ensured with the use of PES timestamps during the packetization process. Any impairment of these time stamps is therefore a source of asynchrony. A non-proper packet-loss concealment strategy may also result in audiovisual asynchrony. An extreme example would consist in not replacing lost audio frames, instead of replacing the missing audio frames by silence or previous audio frames, while using freezing as video packet-loss concealment. The audio would then be shortened and the audiovisual synchronization be lost.

The perception of audiovisual delay has extensively been studied in the literature. A summary has been provided by Kohlrausch and van de Par (2005) and by Hollier et al. (1999). Hollier highlights that the detection of synchronization error depends on the task undertaken by the subject, on the content of the stimulus, and on audio and visual cues. Most of the studies, if not all, agree on the asymmetry of the perception of the audiovisual delay, this delay being more easily detectable when audio is leading. This asymmetry explains why (ITU 1993) recommends for the international exchange of television signals a time difference of maximum 20 ms between audio and video when audio is advanced with respect to the picture and of maximum 40 ms if the sound is delayed with respect to the picture. The asymmetry is assumed to be due to the ecological experience of humans that acoustic information is transmitted less fast by the medium such as air than the visual information, so that it is more plausible that audio is delayed over video than vice versa (Hollier et al. 1999).

The predominance of audio quality and video quality on the overall quality depends on the *content type*, the application and the task/usage, as shown in Jumisko-Pyykkö (2007). Hollier et al. (1999) has studied the impact of video quality on the perceived audio quality, showing that for talking head content, there is a significant change of audio quality when the video quality is changing. This was not the case for content where the video part was an animation and the audio part a narration. This means that for some contents, taking into account the quality impact due to one modality (audio or video) on the quality of the other modality (video or audio) in the modeling of audiovisual quality might improve the quality prediction.

The quality of audio in presence of video, and the quality of video in presence of audio are generally not considered in the models, as they are discussed. This is in agreement with Beerends et al.'s results who showed that considering these two additional interactions did not improve the performance of the model they developed based on their test results.

Finally, ITU (2003b) summarizes which perceptual and cognitive factors have to be addressed in the context of multimodal/audiovisual integration. As perceptual factors, there are the cross-modal influences, e.g., the interactions between different quality levels in the different modalities, the effects of cross-modal error frequency, the auditory-visual error synchrony and the cross-modal masking effects. As cognitive factors, ITU (2003b) lists the cross-modal attention, the task undertaken by the subjects and the prior experience and knowledge of the subjects. Note that in our terminology, perceptual factors are referred to as quality features and cognitive factors are either considered as part of the internal reference or as factors external to the audiovisual signal, impacting the attention of the subject.

2.2.2 From Quality Features to Quality Dimensions

The next step consists in linking the quality features to quality (see Fig. 2.9). It can be observed from the previous section that the list of quality features is long. Therefore, there is a need for identifying the underlying perceptual dimensions that contribute to quality.

The most common approach for identifying the perceptual quality dimensions underlying quality are Multidimensional Scaling (MDS) (Carroll 1972) and Semantic-Differential-based attribute-scaling followed by factor analysis (Martens 2002; Teunissen and Westerink 1996, Hall 2001; Wältermann et al. 2006a, b; Mattila 2001). Other qualitative assessment methods include Open Profiling of Quality (Strohmeier 2010, 2011) or, for a specific mean of identifying attributes, the Repertory Grid Method (Kelly 1955).

The following paragraphs provide an overview of studies identifying perceptual quality dimensions for different modalities (audio, video, audiovisual) and relevant for the IPTV application area. A brief description of the employed quality assessment methods is also provided.

Details on how dimensions can be mapped to quality are provided in Sect. 2.2.3.

In the case of an audio signal, where audio may includes speech, noise, or a combination of these types of signals, the most relevant studies for the IPTV application area are listed in Table 2.3. This table indicates for each study the addressed application, the applied multidimensional assessment technique, the addressed modality, and the perceptual quality dimensions underlying the perceived quality for this application.

Hall in (2001) has conducted a multidimensional scaling (MDS) analysis using encoded speech stimuli. Ten coders were used, and no packet loss was introduced. The MDS operates on subjective ratings of dissimilarity between speech stimuli. The stimuli are presented to the subjects by pair within a pair comparison test, and the subjects have to rate how much similar the pair stimuli are, usually using a 7-point Likert scale labeled from "very similar" to "very dissimilar". This creates a stimulus space in which distances between stimuli correspond to perceptual dissimilarities. Three perceptual quality dimensions resulted from this analysis. For identifying the dimensions, subjects were asked to describe the attributes of the stimuli varying along

Table 2.3 Perceptual Quality Dimensions—Literature overview relevant for the IPTV application area

Application	Multidimensional analysis	Modality	Perceptual quality dimensions	Literatures
Speech coding	MDS	S	Naturalness, noisiness, spectral fullness	Hall (2001)
VoIP/PSTN (NB telephone connections)	MDS and Sem. Dif.	S	(Dis)continuity, noisiness, directness–frequency content/coloration	Wältermann et al. (2006b), (2010)
WB telephone connections	MDS and Sem. Dif.	S	(Dis)continuity, noisiness, distance/coloration, lisping/High-Frequency distortion	Wältermann et al. (2006a), (2010)
NB mobile telecommunications	Sem. Dif.	S	Clean speech: synthetic/natural, dark/bright, smooth/fluctuating/ interrupted, bubbling, noisy. Noisy speech: low/high, synthetic/natural/ boiling, smooth/fluctuating/ interrupted, noisy	Mattila (2001)
HD Cathode Ray Tube (CRT)	MDS	V	Colorness, sharpness	Teunissen et al. (1996)
Interactive multimodel service (PC-based TV phone)	Sem. Dif.	AV	Aesthetic feeling, feeling of activity	Yamagishi and Hayashi (2005)
Audiovisual 3D perception	OPQ	AV	Content, video quality, audio quality	Strohmeier et al. (2010)
Mobile 3DTV	OPQ	AV	Video quality, display and disparity degradations	Strohmeier et al. (2011)

The dimensions contributing the most to the perceived quality are marked in bold; *S* Speech, *V* Video, *AV* AudioVisual, *VoIP* Voice over Internet Protocol, *PSTN* Public Switched Telephone Network, *MDS* Multidimensional Scaling, *Sem. Dif.* Semantic Differential, *NB* NarrowBand, *WB* WideBand

a single dimension. This process resulted in the following naming of the dimensions: *naturalness*, which is the main contribution to the quality and has no readily-apparent physical correlate, *noisiness*, and *spectral fullness*, which are associated with the amount of high- and low-frequency energy respectively.

In their study, Wältermann et al. (2006b) followed two paradigms: multidimensional scaling (Carroll 1972) and semantic differential (Osgood et al. 1957). With the multidimensional scaling approach, no specific cue is given to the subjects on the attributes of the listened stimuli. The method is said "unbiased" in that respect. With the semantic differential approach, subjects have to rate a pre-defined set of attributes using bi-polar scales (e.g., "dark-bright" (Wältermann 2006b). In Wältermann et al.'s work, this pre-defined set has been obtained by asking subjects already familiar with the stimuli to describe them with their own words. The resulting vocabulary set was reduced to a manageable number (13) of antonym pairs, which were used as labels for the bi-polar scales.

As for Hall in (2001), Wältermann (2006b) found three perceptual quality dimensions, in this case for VoIP (Voice over IP) and PSTN (Public Switched Telephone Network) applications. Stimuli rated in this study were degraded using different codecs, packet-loss rates, background noise types and levels, noise suppression algorithms and types of terminal equipment (hands-free sets and telephone handset-type electro-acoustic properties). The dominant dimension, i.e., the dimension which has the highest contribution to the perceived quality, was found to be *continuity*, where the deterioration of the continuity is a result of packet-loss. *Noisiness* is also identified as an important dimension, and is linked with quality elements such as background noise in the speech signal. Finally, *directness/frequency content* as introduced by hands-free terminals is found as third dimension, emphasizing that user-related interface effects should be studied, if not included in speech quality models (Raake 2006b, Sect. 3.4). Note that the dominance of dimensions is also related with the specific choice of test conditions and the strength of the degradation used in the test.

Wältermann et al. have extended their study to WB- (WideBand) transmitted speech in Wältermann (2006a). *Continuity* remains the most important dimension, and *Noisiness* is still one of the dimensions. As new dimensions, *distance* and *lisping* were identified. The latter can be interpreted as an unusual lack of high frequency components.

Three of the dimensions presented in Wältermann et al.'s and Hall's work correspond to the dimensions Mattila found (Mattila 2001) for the narrowband mobile telecommunications: continuity, which is named *smooth/fluctuating/interrupted* in Mattila (2001), noisiness (*noisy*) and naturalness (*synthetic/natural*). Two additional dimensions are found for clean speech (*dark/ bright*, *bubbling*), and one more dimension for noisy speech (*low/high*). Note that the stimuli were degraded with typical mobile communication degradations: car cabin noise, packet loss, coding, speech enhancement, tandem connections of speech coders, etc.

For a more exhaustive list of multidimensional analysis of speech transmission, see Raake (2006b), Sect. 3.4, and (Côté 2010), Sect. 1.4.

In the case of video, Teunissen et al. (1996) conducted a study using six TeleVision (TV) sets (CRT) varying in spatial resolution, color reproduction, peak luminance, and luminance contrast. Video stimuli were presented to the subjects in two ambient illumination environments. In an attribute scaling table, eight quality features were found to be relevant for the application (color naturalness, sharpness, darkness, brightness, contrast, flicker, smear/geometrical distortion). The authors identified that *color naturalness* (which is affected by color rendering) was the most important factor, followed by *perceived sharpness*. They further conclude that the combined scores for sharpness and naturalness give a good prediction of overall perceived quality. They observed that the correlation between the sharpness cluster and quality scores is higher (0.83) than between the naturalness and the quality scores (0.69), but that a difference in color (RGB) balance (color naturalness) affects the perceived quality more than a difference in resolution (sharpness).

Yamagishi et al. in (2005) addressed the audiovisual case for interactive multimedia service such as video-telephony. They found that two perceptual dimensions are contributing to the perceived quality: *aesthetic feeling* and *feeling of activity*. The aesthetic feeling is linked to audio and video packet loss, and video bitrate. This dimension is related to attribute pairs such as quiet/clamorous, clear/cloudy or beautiful/dirty. Feeling of activity is related with one-way transmission delay and video frame rate. Related attributes are, for instance, dynamic/static, slow/fast, or light/heavy.

Strohmeier et al. in (2010, 2011) adapted the Free Choice Profiling (FCP) method commonly used in the sensory evaluation of food (Lawless and Heymann 1999) to the analysis of the perceptual dimensions underlying the quality for 3D audiovisual applications such as Mobile 3DTV. This new method is referred to as Open Profiling of Quality (OPQ) and consists in three sessions. The first session is an ACR (Absolute Category Rating, see Sect. 2.3.1) test in which subjects rate the quality of the stimuli (note that OPQ is not restricted to quality, and the acceptance or the overall satisfaction could be rated instead of quality). In Strohmeier et al. (2010, 2011), the subjects filled out a demographic questionnaire after the first session. In the second session, subjects are told to think about quality features they have used to evaluate overall quality in the first session. Features which are not unique or cannot be defined are excluded. The resulting list of attributes is written on a scoring card, with a 10 cm continuous scale for each attribute, with a "min" (minimum sensation) and a "max" (maximum sensation) label at the extreme ends of the scale. In the third session, each subject rates the stimuli using all scales on their scoring card.

In Strohmeier et al. (2010), two 3D audio and video contents corresponding to a virtual classroom and a virtual living room were used. In the virtual classroom, the audio was speech (from a male speaker) and the sound source was a manikin. In the virtual living room, the sound source was represented by a laptop playing drum and bass music. Two acoustic models based on acoustic room simulation, corresponding to the virtual rooms, were produced. The simulated user's movement in the room was a straight approach towards the sound source, and turns to the left and to the right. The contents were played out on a 15 inch stereoscopic display and on 4.0 loudspeakers. The visual and auditory perception of depth were varied by switching the display from mono to stereoscopic view and by interchanging the room acoustic models between rooms. The results of the OPQ tests were analysed with the Generalized Procrustes Analysis (GPA) and three dimensions were found: a *content* dimension, linked for instance to the attributes "tidy" and "comic like", a *video quality* dimension, linked to the perception of depth in video and to the related degradations, and an *audio quality* dimension, linked to attributes describing the change of room acoustics and therefore perception of depth ("reverberation", "echo"), or to affective attributes such as "thrill", "dominant" and "anxiety".

Another important output of the study is the observed differences between subjects: The attributes they use do not correlate with the same dimensions, and only few subjects use two or even three dimensions for deriving their quality attributes.

In Strohmeier et al. (2011), six 3D audiovisual contents were used and the video parts were H.264 encoded with three levels of Quantization Parameter (QP), from low to high. The audio was encoded with the AAC encoder by Nero. The audio bitrate was fixed, and therefore the audio quality did not vary. The OPQ test was conducted in two environments: a controlled laboratory and a café. The same perceptual dimensions were obtained in the two environments, showing that the OPQ method—and therefore the obtained perceptual dimensions—were robust to the context of use. Two perceptual dimensions were derived: a *video quality* related one, linked to the attributes "blocky" or "sharp", and a *display and disparity degradation* related dimension, linked to the cross-talk effect and to attributes such as "perceivable as one image", "double effect", "dark", "annoying". Although the same dimensions were obtained in both environments, some differences on the loadings of attributes on the different dimensions exist. In particular, the number of attributes showing high correlation with the first dimension is lower in the café environment. Moreover, there is a higher sensitivity to the second dimension in the café than in the lab, confirming the results of other studies in which ease of use and viewing comfort were reported as important components of quality in the context of use.

The impact of the context on the perceptual quality dimensions has also been studied by Wältermann in Wältermann et al. (2010). However, "context" does not refer here to the user environment but rather to the quality range and degradation types covered by the stimuli used in the study (see Sect. 2.3.2 for a more detailed description of the "context effect", in particular the "corpus effect"). In his work, Wältermann re-analyses the data obtained in Wältermann (2006a, b), and in which several multidimensional analysis campaigns had been conducted for two speech telephone campaigns: a traditional Narrowband (NB) one, and a mixed Narrowband/Wideband (WB) one, as often encountered in Voice-over-IP services. The same three dimensions were obtained in both scenarios: *Discontinuity*, *noisiness*, and *coloration*. In both cases, the *discontinuity* is the dominant dimension, followed by *noisiness* and *coloration*. In the mixed NB/WB scenario, a fourth dimension *high-frequency distortion*, specific to WB, was identified. This study confirms the validity of the multidimensional analysis approach, it especially emphasizes the robustness of the obtained dimensions towards new types of degradations and different quality ranges. The authors highlight that the weighting factors of the dimensions underlying quality (not the dimensions themselves) may change with different contexts, where contexts here refer to the quality ranges and degradation types, or if certain impairments become more dominant.

By mixing quantitative and qualitative tests, the method used in Jumisko-Pyykkö (2007) is close to the one use in Strohmeier et al. (2010), but semi-structured interviews were conducted instead of Free Choice Profiling. Five major categories were derived from the large amount of interviews (60 persons participated in the test): the content factors ("News", "Animation", etc.), the usage factors ("purpose of use", "equipment", etc.), the audio quality ("disturbing inferiority"/ "erroneousness", "background sound", "echo"), the video quality ("details", "disturbing inferiority"/"erroneousness", "blurriness", etc.), and the audiovisual quality ("audio quality more important than video", "synchronism"). For examining the dependencies between the factors, Bayesian classification modeling was applied. Both the classification of the quality factors per *content type* or based on the category *disturbing inferiority/erroneousness* resulted in accurate models. When classifying the factors per content type, it was observed that some contents are related with more vision-related factors, other are related with both audio and video specific factors.

2.2.3 From Quality Dimensions to Quality

This section relates to the topic of preference mapping and corresponds to the last step of the full path from quality elements to quality (see Fig. 2.9). As described in Carroll (1972), the perceptual quality dimensions can be either of the vector- or of the ideal-point-type. With vector-type dimensions such as "Naturalness", the more of that dimension, the better (the quality). In contrast, dimensions like "loudness" or "brightness" have an ideal point which yields the highest quality. Following Carroll's work, vector-type dimensions can be linearly combined for estimating the perceived quality, while quadratic regression is used for ideal point-type dimensions. A combined approach has been proposed in Wältermann (2012).

The work of Allnatt (1983) is worth mentioning in this context. Indeed, he assumes that certain kind of impairment factors may be considered as additive on an appropriate (perceptual) quality-rating scale. Impairment factors here are the quality-related counterpart of technical degradations, i.e., the transformation of technical degradations onto a perceptual quality scale in terms of impairments. These impairment factors are assumed to be equivalent to the perceptual quality dimensions presented in the previous section. The impairment-factor-based approach can therefore be considered as a vector model. This concept initially developed for TV-type applications (Allnatt 1983) has been adopted for the development of the so-called E-model, a planning model for speech communication services (ITU 2005). The linear mapping was re-analysed in Wältermann (2012), where the impairment factors have been replaced by the corresponding perceptual dimensions. In the respective mapping found by Wältermann in his optimization for different test campaigns, a Euclidean distance type model is used for the impairments. This implies an ideal point for the case of "no impairment", which corresponds to the view of a threshold of no or zero perceivable impairment.

Note that a multiplicative approach may also be followed instead of the additive approach, as in Yamagishi and Hayashi (2008), where Yamagishi et al. are combining impairment factors related to compression artifacts and to transmission errors. Finally, a mix of the additive and multiplicative approaches can be used, as commonly done for modeling the audiovisual quality. The discussion on audiovisual quality modeling is reserved for Chap. 6.

It should be observed that the type of attributes and dimensions described in Jumisko-Pyykkö (2007), Strohmeier et al. (2010, 2011) is differing from the ones reported in the other studies listed in Table 2.3. In particular, Jumisko-Pyykkö et al. are referring to content- and usage-related attributes in addition to the quality-related ones, and Strohmeier et al. have found a "content" dimension in Strohmeier et al. (2010). This appearance of new types of attributes and dimensions is partially linked to the employed method (OPQ, semi-structure interviews), but also to the task given to the subjects as well as the additional aspects, such as content and context of use, being considered. In previous studies, the attention of the subjects was focused on the stimulus characteristics (including degradations), probably because they were asked to describe the stimuli, while in Jumisko-Pyykkö (2007), Strohmeier et al. (2010, 2011), subjects are asked to report which criteria/factors they use for judging the quality (or acceptance or satisfaction) of the stimuli. Following the example of Strohmeier's study, the content dimension will probably not be used similarly as the audio and video quality dimensions in the modeling. These latter two dimensions could indeed be combined into the vector model, while content and usage (context) will probably act instead in terms of weighting of the different perceptual quality dimensions. In other words, for a given content and context of use, the first dimension may have more impact on the overall quality than for another content in another context. This relates to the open topic of overall QoE instead of quality modeling and will not be addressed further in this work.

All quality-related concepts have now been identified and defined. As mentioned at the beginning of this chapter, next steps consist in selecting appropriate test methods and rating scales for conducting the subjective quality tests which will serve as basis for the model development presented in this work. These topics are addressed in the following section.

2.3 Subjective Quality Assessment

The design of subjective tests are crucial since quality models are developed based on the results of these tests. Any bias in the subjective tests may decrease the performance of the model, or lead to a reduction in model validity.

Perceptive audio, speech, video and audiovisual quality tests are commonly conducted using standardized ITU (or EBU) methods. One of the main advantages of standardized methods is to make subjective tests reproducible and comparable between different laboratories and even across test runs within a given laboratory.

Note that in the following, the terms "subjects" and "observers" will be used for designating test participants, regardless of whether the tests are audio, video, or audiovisual.

Commonly used and established recommendations are P.800 (ITU 1996b) for speech, BS.1116 and BS.1534 (ITU 1997, 2003a) for audio, BT.500, BT.710, P.910,

BT.1788 (EBU SAMVIQ) (ITU 1998a, 1999, 2007, 2009; EBU Report 2003) for video, and P.911 and P.920 (ITU 1998b, 2000) for audiovisual quality tests.

These documents provide recommendations on the listening and viewing environment, audio terminal and display set up, on the test material and test design to be used in the test, on the amount and characteristics (expert/non-expert, gender, age) of the subjects to be recruited. They also provide a selection of test methods and rating scales, and specify the task for which each method is the most appropriate (see for instance Table 2 in (ITU 2009) for TV pictures). Some guidelines on the data analysis are also given.

It is not the scope of this book to provide an exhaustive overview of all existing test methods for audio/speech, video, and audiovisual quality assessments. However, some papers are reviewed in Sects. 2.3.1 and 2.3.2 for highlighting important points to be considered when selecting a test method and the associated rating scale. For an overview of audio, speech, and video quality methods, the reader is referred to Bech and Zacharov (2006, Fig. 1.3, 1.4 and Annex A), Möller (2000, Chap. 4) and Wu and Rao (2006b, Chap. 4) respectively. These books also provide an overview of rating scales and the related literature. On this latter topic (the rating scales), the interested reader may also refer to the work conducted by Zielinski et al. in (2008) and Huynh-Thu et al. in (2011).

2.3.1 Standardized Test Methods

The standardized test methods differ in many aspects (Huynh-Thu et al. 2011): (a) their presentation pattern, (b) their rating type and (c) the type of rating scale. The type of rating scale is covered in the next subsection.

The presentation pattern refers to the number of stimuli presented to the subject per trial/presentation (single-stimulus, double-stimulus, multi-stimulus, pair comparison). In the case of double-stimulus presentation, one of the stimuli is the reference signal, i.e., the (non-degraded) source signal. This reference can be explicitly presented to the subject, i.e., the subject knows which of the two presented signals is the reference. Otherwise, the reference is said to be "hidden" to the subjects. Both the reference (hidden or not, depending on the method) and the degraded signals may be presented twice per presentation slot, as in the Double Stimulus Continuous Quality Scale (DSCQS) (ITU 2009) method. In the case of multiple-stimulus presentation, as in the MUltiple Stimuli with Hidden Reference and Anchor (MUSHRA) (ITU 2003a) or the Subjective Assessment Method for VIdeo Quality (SAMVIQ) (ITU 2007; EBU Report 2003) tests, there is usually both a hidden and an explicit reference in the same presentation slot. The subjects can also re-view/re-listen to the stimuli if they want to.

When the rating type is "retrospective" (Hands and Avons 2001), the subject rates the single-/multiple-stimulus at the end of the presentation slot, i.e., after the subject has seen the whole stimulus. Note that the duration of each stimulus is typically 5–8 s for speech and 5–16 s for video or audiovisual tests. The stimuli should not be longer in order to avoid the "primary" and the "recency" effects, which are cognitive biases resulting in a subject recalling better events (e.g., audiovisual artifacts) occurring at the beginning and at the end of the stimulus (Aldridge et al. 1998; Hands and Avons 2001. Due to this bias, these latter events have more weights in the retrospective quality rating of the subject.

Most of the test methods use "retrospective" ratings. With "continuous" rating, as with the Single Stimulus ContinuousQuality Evaluation (SSCQE) method (ITU 2009), the subject is continuously rating the observed stimulus. Ratings are usually automatically stored at a frequency of two ratings per second. With this rating type, the duration of the stimuli is of several minutes (typically 20–30 min), making the tests closer to a more natural listening/viewing situation.

In the context of IPTV, the methods listed in Table 2.4 are of interest. The table indicates as well, for each method, which rating scales are used (see also Fig. 2.17).

Four criteria are commonly used for comparing subjective quality test methods (Huynh-Thu et al. 2011): the accuracy, the precision, the efficiency, and the repeatability (stability/reliability). The accuracy is the distance between the measured mean value of the ratings and the true mean value that would have been obtained with the entire population (instead of a small sample (our subjects) of this population). In practice, the accuracy cannot be assessed. Instead, the mean values of different subjective test methods are compared, and from the results the most accurate method is identified. The precision of a test method is obtained by computing the confidence interval associated with the mean ratings. The smaller the confidence interval, the higher the precision of the method. The efficiency of the method will be represented here by a trade-off between the number of stimuli rated by the subjects, the number of subjects required for obtaining a low confidence interval, and the test duration. Also interesting to check is the ability of a method to reproduce statistically identical results if the experiment is repeated (in the same lab or in different labs), i.e. its repeatability.

Most of these criteria have been used in the following studies for comparing subjective test methods.

The most commonly used test method formerly was the DSCQS method (ITU 2009). Since stimuli are presented in pairs (the hidden reference and the processed video), this method is sensitive to small differences. It also limits context effects, which occur when subjective ratings are influenced by the severity and ordering of impairments within the test session (see Sect. 2.3.2).

Some methods are more suitable for assessing time-varying video quality, such as the SSCQE where the subject is continuously rating (e.g., 2 ratings per second) long sequences (20–30 min) (Zou and Corriveau 1996). This method is also more representative of the quality monitoring situation, as mentioned in Pinson and Wolf (2003) and is more efficient since, compared to non-continuous-evaluation methods, more (concatenated) clips can be rated for the same viewing time. However, with one rating every half a second, data interpretation has its own difficulties.

Pinson and Wolf (2003) in (2003) conducted a SSCQE test on 45 s "super-scenes", composed of concatenated independent 8–9 s sequences. They show that, if the hidden reference removal technique is applied, the rating sampled at the end of each

Table 2.4 Standardized test methods relevant in the context of IPTV [see also Bech and Zacharov (2006) and Wu and Rao (2006b, Chap. 4), for an overview of audio and video quality test methods]

Modality	Reference	Description
Speech	ITU-T P.800 ACR	No reference; single stimulus; non-expert subjects; 5-point categorical quality scale (see Fig. 2.17a); large impairments; good reliability; wide quality range; up to WideBand speech
Audio	ITU-R BS.1116	Hidden and known references; triple-stimulus; expert subjects; impairment scale (see Fig. 2.17b); for small impairments; small quality differences; excellent reliability; mainly used for audio, recently used for Full Band speech codec quality
Audio/speech	ITU-R BS.1534 (MUSHRA)	Hidden and known reference; multiple stimulus; trained subjects; continuous quality scale with quality labels and numbers (see Fig. 2.17c); good reliability; intermediate quality levels; used in most audio codec quality tests
Video	ITU-R BT.500 DSCQS	Hidden reference; double-stimulus, double presentation; non-expert subjects; continuous scale with quality labels but no numbers (see Fig. 2.17d)
Video	ITU-R BT.500 SSCQE	(Hidden reference); non-expert subjects; continuous scale with quality labels but no numbers (see Fig. 2.17e); long stimuli (20–30 min)
Video	ITU-R BT.500 DSIS	Explicit reference; double-stimulus, single or double-presentation; impairment scale (see Fig. 2.17f)
Video	ITU-T P.910 ACR	See ITU-T P.800
Video	ITU-R BT.1788 (EBU SAMVIQ)	See ITU-R BS.1534
Audiovideo	ITU-T P.911 ACR	See ITU-T P.800.

ACR Absolute category rating, *MUSHRA* MUltiple stimuli with hidden reference and anchor, *DSCQS* Double stimulus continuous quality scale, *SSCQE* Single stimulus continuous quality evaluation, *DSIS* Double stimulus impairment scale, *SAMVIQ* Subjective assessment methodology for video quality. Note that when not otherwise indicated, stimulus duration is of approx. 5–10 s

8–9 s video clip of the 45 s “super-scene” is highly correlated with quality scores obtained on the same 9 s video clips but using the DSCQS method.

With the emergence of new multimedia technologies, Absolute Category Rating (ACR) (ITU 1999) has progressively replaced the DSCQS. ACR was less mature than DSCQS but four times faster, since a single stimulus is presented once with ACR, while a double stimulus is presented twice with DSCQS. One of the results in Huynh-Thu and Ghanbari (2005) shows that, for low-bitrate and low-resolution video, the Mean Opinion Score (MOS) scores (scores averaged per file or per test condition on the 5-point ACR scale) given by subjects in DSCQS are highly correlated with the MOS scores given in ACR with Hidden Reference. This is a really positive result since ACR allows testing many more video files in one test.

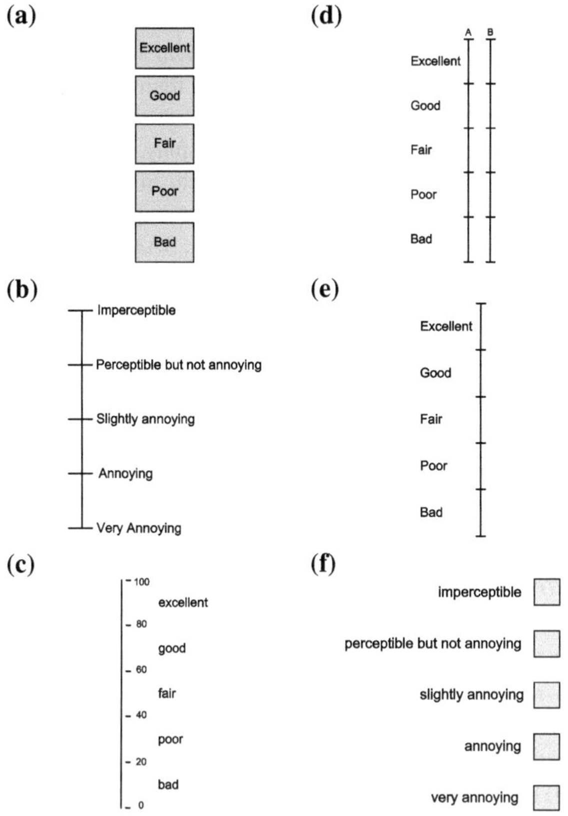

Fig. 2.17 Rating scales used in standardized test methods (see Table 2.4). **a** 5-point categorical quality scale [ACR (ITU 1996b, 1999, 2009)]. **b** Impairment scale used in BS-1116 (ITU 1997). **c** Quality scale used in MUSHRA/SAMVIQ (ITU 2003a; ITU 2007). **d** Quality scale used in DSCQS (ITU 2009). **e** Quality scale used in SSCQE (ITU 2009). **f** Impairment scale used in DSIS (ITU 2009)

The ACR method also has the advantage of showing a good inter-lab stability. For instance, in Huynh-Thu (2006), results coming from the same ACR test conducted in two different labs correlate with a Pearson correlation coefficient of 0.98. Similar performance was found in Brotherton (2005).

The ACR method has been recently compared to the SAMVIQ method (Brotherton 2005; Rouse et al. 2010). SAMVIQ is an EBUmethod specifically designed for assessing the perceptual quality of video delivered as part of multimedia services. SAMVIQ is similar to the MUSHRA method but for video, as indicated in Table 2.4. It is a multiple stimulus method. Several stimuli corresponding to the same content are presented to the subject, and each stimulus should be rated using an 100-point scale with the labels "Excellent","Good","Fair","Poor", and "Bad" located to 90, 70, 50, 30, and 10 on the 100-point scale (see Fig. 2.17c). Explicit and implicit references are part of the multiple stimulus, and the subject can re-view each stimulus as many times as she/he wants. According to Brotherton (2005), SAMVIQ and ACR are highly correlated, provided that there is a sufficient variation in quality across the set of stimuli. However, SAMVIQ is more critical than ACR, i.e., it allows a better distinction between stimuli (Rouse 2010; Brotherton et al. 2005). This may be an advantage for applications with high resolution display, but may be unnecessary for applications using low resolution display for which, in real life viewing situations, small differences between stimuli would not be detected. Moreover, by presenting in a same trial multiple stimuli from the same content, SAMVIQ encourages observers to develop separate quality scales for stimuli generated from a particular reference content. As a result, the perceived quality scores for stimuli generated from different reference content may be misaligned (Brotherton 2005). Finally, and this is its main advantage, SAMVIQ can be conducted with less subjects than ACR, typically 15 instead of 24. Indeed, based on the analysis of the confidence intervals, Rouse (2010) concluded that SAMVIQ scores have greater precision than ACR scores for the same number of observers. This compensates for the fact that, by design, less stimuli can be rated with SAMVIQ than with ACR.

2.3.2 Standardized Rating Scales

Rating scales differ in many terms: discrete versus continuous, labeled versus non-labeled, quality-related- (bad to excellent) versus degradation-related (very annoying to imperceptible) labels, number of labels, presence of numbers or not. Some of these aspects, such as the selection of the labels and their location on the scale, are impacting the type of the scale. Four types of rating scales have been identified by Stevens in (1946): "nominal", "ordinal", "interval", and "ratio". "Nominal" scales allow allocating membership to an item and checking if two items are equivalent (i.e., from the same group) or not. As indicated by their name, "ordinal" scales are used for ranking ("ordering") items. "Interval" scales are also ordering the items but, in addition, the distance between items is known. "Ratio" scales are interval scales with a true point zero. The selection of the rating scale is motivated by the type of results to be obtained and the type of statistics to be conducted on the results. If the target is to rank different codecs, an ordinal scale (codec A is better than codec B) is sufficient. But for knowing how much better codec A is compared to codec B, an interval scale is more appropriate. With an interval scale, the distance between the items is known, and statistics such as the arithmetic mean, the variance and the Pearson correlation coefficient can be computed. More details can be found in Bech and Zacharov (2006, Chap. 4).

In quality testing, the categorical scale such as the ACR 5-point scale (see Fig. 2.17a) is used in terms of interval scale, even though several drawbacks are indicating that these scales are closer to the ordinal scale than the interval scale. For instance, Jones and McManus showed in Jones and McManus (1986) that the labels of the five-point ACR scale are not equidistant on a perceptual continuum, and the distance between labels vary from one country/culture to another one. This problem can be circumvented by using additional numeric labels (Gleiss 1987) and/or a graphical scale representation underlying the interval-character of the scale, see also (Huynh-Thu et al. 2011). The five-point ACR scale also suffers from "saturation effects" (Möller 2000). This phenomenon reflects the fact that once a subject has rated one stimulus as for instance "Excellent", she/he cannot rate the following stimulus better than "Excellent", even if it is perceived as such. This effect may be limited by properly training the subjects before the tests. During this training phase, the subjects should be presented stimuli covering the whole quality range, all contents and degradation types to be judged later in the test. This saturation effect may also be avoided by extending the extreme end points of the scale, as in the "Eleven-grade numerical quality scale" of the ITU-T P.910 Recommendations (ITU 1999, Fig. B.2/P.910), or as in the 7-point scale proposed by Bodden and Jekosch in Bodden and Jekosch (1996). This latter improvement of the scale may also help in limiting the tendency the subjects have to avoid the extreme ends of the scale.

The ACR scale is further criticized for its lack of sensitivity (Möller 2000). Indeed, it is questionable if more than five categories could convey a higher level of information provided by the subjects, or if this would introduce more response errors. Based on a literature review and an own study, Cox concluded in (1980) that there is no apparent advantage in using a number of scale categories larger than nine, while a very small number of categories (less than five or six) could produce a loss in accuracy.

For a wider literature review on rating scales, the reader is referred to Huynh-Thu et al. (2011) and Möller (2000).

Zielinski et al. (2008) provide a list of biases encountered in listening quality tests, indicating at which stage of the test design the biases occur, from the selection of stimuli to be listened to by the subjects to the interface they use for rating their opinion (paper sheet, computer-interface, etc.). These biases are valid for video and audiovisual quality tests as well. They can be further classified into "subject effect", "response-mapping-scale effect", and "context effect".

The "subject effect" is detailed by Zielinsky et al. in (2008). It refers to the expectation and characteristics of the subject which may impact his/her judgement. For instance, this judgement may be affected by the brand of the devices used in the test, by the mood and personality of the subject, or by the situational context in which the subject is (in a test laboratory or in a living environment such as a café).

The "response-mapping-scale effect" is broadly covered by Zielinsky et al. and Möller in (2008, 2000). Basically, this effect refers to the biases which occur when the subject maps his/her opinion to the scale given by the experimenters.

The "context effect" includes the "order-effect", the "distribution-effect", and the "anchor-effect" (Möller 2000). The "distribution-effect" further includes the "corpus-effect". Broadely, those effects describe biases introduced by the selection of stimuli to be rated, in particular the quality range and degradation types they cover. For instance, in Möller (2006) and Côté (2008), the authors show that the introduction of bad-quality conditions in a test leads to higher ratings of all conditions in the test, and vice versa. This is referred to as the "positive" and "negative" corpus effect. The authors also observe that transmission errors and background noise are the dominant degradations and they likely draw the attention away from speech codec distortions. These effects may be overcome by using anchors – or reference units—in the subjective tests (ITU 1996a, d). Such anchors, also referred to as common set in the literature, were in particular used in competitions conducted for video quality model standardization in the Video Quality Expert Group (VQEG) (Pinson and Wolf 2008), or in the ITU-T campaigns for developing the Perceptual Objective Listening Quality Assessment (P.OLQA), Parametric Non-intrusive Assessment of audiovisual Media Streaming quality (P.NAMS) and Parametric Non-intrusive Bitstream Assessment of video Media Streaming quality (P.NBAMS). Additional details can be found in Garcia (2010a). Note that the range of quality and degradations covered by the anchors—or common set—should be ecologically valid, i.e., they should be representative of the range of quality and degradations encountered in the "real" implementation of the application.

The "order-effect" mainly describes that the judgement of the current test stimuli is influenced by the quality of the precedent test stimuli. This effect may easily be overcome by having different presentation orders per subject.

The "anchor-effect" refers to the comparison the subject makes between the test stimuli and his/her internal reference, this reference being influenced by his/her previous experience with the stimuli, but also by the first stimulus he/she has to judge. This effect may be attenuated by having several subjects participating in the test and by familiarizing the subjects during a training period with the rating scale and with the range of quality and degradations used in the test.

2.4 Quality Models

Quality models, also referred to as "instrumental measurements" or "objective models", estimate the quality of multimedia systems as perceived by the user and based on signal measurements (or assumptions when the signal is not available, as in a network planning phase). Which type of model should be selected or developed depends on the targeted application, on the type of the available input signal(s), and on many other aspects which are described in the following section. Models also differ in terms of the modeling method employed during the model development. An overview of modeling strategies is provided in Sect. 2.4.2. Performance metrics commonly used for evaluating and comparing quality models will also be reviewed, see Sect. 2.4.3.

2.4.1 Quality Model Types

Multimedia quality models can be classified according to the following categories (Winkler and Mohandas 2008; Raake et al. 2008, 2011; Takahashi et al. 2008):

- The amount of information needed from the original signal (Winkler and Mohandas 2008; Raake et al. 2008, 2011):

  - Full Reference (FR): A reference signal is required. FR-models compare the processed signal to the original (reference, source) signal,
  - Reduced-Reference (RR): Partial information extracted from the source signal is required. This partial information is compared to the partial information extracted from the processed signal,
  - No-Reference (NR): No reference signal is required, and no comparison between the processed and original signal is done.

- The type of input the model has access to:
  - Planning models: The model is used during the network planning phase. As a consequence, the model has only access to side information such as the codec type and profile, the expected bitrate and packet-loss rate,
  - packet-header-based models: Information is extracted from the stream during monitoring, based on packet-headers,
  - bitstream-based models: Bitstream-level information is required. In Raake (2011), the bitstream-based models are further split into two categories: mode 1 and mode 2. In mode 1, the bitstream is not fully decoded, i.e., no pixel-information or audio signal is available after decoding, while in mode 2 a full-decoding is allowed,
  - signal/media-based: The decoded image (pixel-information) or the audio signal is required,
  - hybrid model: This model corresponds to a mix between the media-based model and either the packet-header or bitstream-based model.
- The type of application the model has been designed for:
  - Codec or service component testing,
  - network planning: The model or measurement system is used before the implementation of the network in order to plan the best possible implementation,
  - quality (or "service") monitoring: the model is used during service operation.
- The target service (Raake et al. 2008):
  - The service type the model is covering, e.g., IPTV, mobile TV, or other,
  - the video resolution the model is addressing, e.g., HD, SD, …
- The model output (Raake et al. 2008):
  - Ranking of coders,
  - overall integral quality,
  - diagnostic information, e.g. related to the perceptual quality dimensions underlying the perceived quality (see Sect. 2.2.2), or to the quality element(s) causing the given quality estimate.

Takahashi et al. (2008) and Raake et al. (2008) provide an overview of the standardization activities completed and on-going for each model types.

In the following, planning-, packet-header- and bitstream-based models will be referred to as *parameter-based* models.

In Winkler and Mohandas (2008), Winkler et al. further classify the video signal-based models into two categories: the “data metrics” and the “pictures metrics”. “Data metrics” are targeting measuring signal fidelity without considering its content, while “picture metrics” consider the visual information of the video data. Further classification can be done based on the applied modeling approach, e.g. vision modeling versus an engineering approach, where the vision modeling approach is targeting models of the human visual system, while the engineering approach analyses the video data for extracting and analysing features relevant for quality, without explicit perception model.

The type of model input and the targeted application type are closely related. For instance, for codec testing, the task can be completed offline. As a consequence, signal-based FR models, which require more processing resources and time than all other model types, but are also potentially more accurate since they have access to more information, can be used. For the quality monitoring of mobile applications such as MoTV, the models may be implemented on the mobile terminals themselves. As a consequence, the models do not have access to the source signal and they should not be computationally demanding. The same condition applies to model clients implemented on set-top-boxes (STBs) or internet access devices (IADs) in IPTV applications; here, too, processing resources are limited, Hence, packet-header or bitstream-based models (mode 1) are the most appropriate candidates in these cases.

2.4.2 *Model Development*

Considering that the model type has been selected, the model development can be split into three main stages (see Fig. 2.18): the selection of the variables on which the model is trained, the modeling itself, and the evaluation of the model. Depending on the evaluation results, the whole procedure may be iterated multiple times. The “modeling” stage further comprises important steps such as the selection of the modeling method, e.g., multiple regression analysis, and the selection of the functions for combining the selected variables and fitting them to quality. As shown in the following, temporal pooling is part of these combination and fitting processes. The variable selection and the modeling steps are described in this subsection, the evaluation step in the following one.

2.4.2.1 Feature-Based Modeling Strategy

One modeling strategy, that will be referred to as “feature-based modeling strategy” in the following, is valid for media-based and especially FR models. It consists in comparing the degraded and reference (pixel-based) signals frame-by-frame, or even block-by-block, based on “video-features” such as luminance, contrast, and structure (Wang et al. 2004), or on characteristics related with the Human Visual System (HVS), such as spatio-temporal masking and the contrast sensitivity function (Winkler 1999). Once video- or HVS-based features have been derived per image

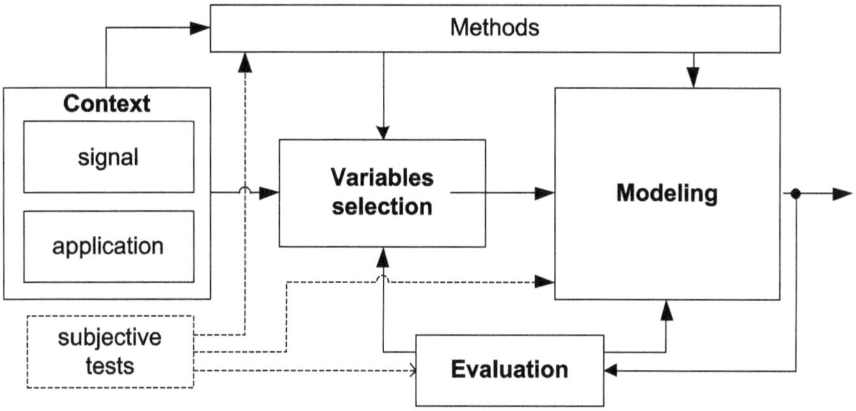

Fig. 2.18 Model development framework

block or frame, they are combined and/or pooled for obtaining a metric for the whole video sequence. The well-known Peak-Signal-to-Noise Ratio (PSNR) fits this modeling strategy, but no features are extracted from the video signal, and instead pixel-wise comparison is applied using the Mean Squared Error (MSE) between test and references pixels. More information on video quality models following the "feature-based modeling strategy" can be found in Winkler (2005) and Chikkerur et al. (2011).

2.4.2.2 Degradation-Based Modeling Strategy

Another approach for quality modeling, that will be named "degradation-based modeling strategy" in the following, consists in finding a metric for each perceptual quality dimension as defined in Sect. 2.2.2, before combining them to an integral quality estimation. As shown in Sect. 2.2.2, perceptual quality dimensions may be obtained by carrying out a multidimensional analysis. However, multidimensional analysis is expensive and time-consuming, so that not all researchers make the effort to conduct such studies before running standard quality tests and developing quality models. In most cases, the degradations considered as the components of quality are not dimensions analytically found in multidimensional tests, but assumed dimensions identified by experts. For instance, Farias et al. predict the perceived annoyance by combining blocking, blurriness, and noisiness metrics (Farias and Mitra 2005).

The variables which will be used as input to the quality impairment terms (representing the perceptual quality dimensions), and consequently to the integral model, are therefore selected based on their relevance and contribution to the perceptual quality degradations. Note that the addressed degradations highly depend on the targeted application. This is the reason why "application" is one input of the "variable selection" box shown in Fig. 2.18.

When a large amount of variables are good candidates for input to the model, a factor analysis such as Principal Component Analysis (PCA) may be necessary for selecting the optimal set of inputs to the model.

As an example of media-based metrics following the "degradation-based modeling strategy", the Video Quality Model (VQM) from NTIA (Pinson and Wolf 2004) is explicitly modeling quality based on a weighted linear combination of blurring, block distortion, jerky/unnatural motion, noise in luminance and chrominance channels, and error blocks (resulting from, e.g., transmission errors). Yamagishi and Hayashi (2009) is one example among many others of parameter-based video quality model following the "degradation-based model strategy". Their model combines a video-compression impairment module with a video-transmission-error impairment module.

2.4.2.3 Temporal Pooling

Yamagishi et al.'s model (2009) differs from the previously mentioned studies and especially media-based metrics in different aspects. First, the "selected variables" (the model inputs) are measured over the whole sequence, e.g., bitrate and packet-loss-rate. In other terms, instead of a applying a temporal pooling on the predicted quality per frame, as in most of the media-based quality models, it is done on the variable itself. The assumption behind this is that the observer of the video sequence does not form a quality judgement per frame and then mentally averages it, but instead perceives each (possibly time-varying) degradation as a whole entity and forms a quality judgement on a per-sequence basis.

Temporal pooling may just be performed by calculating the mean of the per-frame feature or quality, as in Wang et al. (2004). More sophisticated temporal pooling methods exist, such as the Minkowski summation, the exponentially-weighted Minkowski, and the mean of the last F frames, as described in Rimac-Drlje et al. (2009). This latter study, which compares various temporal pooling methods for different video quality models, highlights that temporal pooling methods should reflect both the influence of the worst quality part of the sequence and of the recency effect. This is for instance done in the temporal pooling metrics proposed for speech by Weiss et al. in (2009), by Rosenbluth in (1998) and Gros and Chateau in Gros and Chateau (2001).

2.4.2.4 On the Use of Subjective Test Results

Another aspect specific to the parameter-based quality models is the use of subjective test results during the whole modeling process. Here, the modeling functions are obtained by mapping the selected variables to the subjective quality ratings.[3]

[3] Quality ratings estimated by (preferably) FR quality models could be used as complement, but the model developed based on these ratings then mimics shortcomings of the FR model.

The mapping is commonly done by applying a regression analysis using the subjective test results as target values. The available variables are limited to the number of parameters which could be varied during the subjective tests. As described earlier, regression is often done in terms of the “degradation-based modeling strategy” using the quality contribution terms related with different types of variables.

The rating scale used during the subjective tests is also impacting the modeling method. As previously mentioned, rating scales are sources of biases in the quality ratings. In particular, the use of a categorical scale may result in a saturation effect at the extreme ends of the scale, leading to non-normal distribution of the ratings. In the case of regression analysis, the assumption of normal distribution and uniformity of variances of the residuals may not be fulfilled, which are usually necessary for running the regression analysis.

Note that when several degradations or characteristics co-exist, the model developer has to decide how these degradations should be combined for estimating the integral perceived quality. This relates to the topic of “preference mapping” which has been covered in Sect. 2.2.3.

Factors impacting the model development have now been presented. At this stage, the selection of the most appropriate variable and model has to be adjusted in an iterative process for predicting quality values as close by as possible matching the perceived quality ratings obtained in the underlying tests.

2.4.3 Model Evaluation

The performance of a quality model is computed by comparing the quality scores estimated by the quality model to the quality scores obtained from subjective tests. Ideally, this comparison should be done with unknown subjective test data, i.e., subjective scores which have not been used for training the model. When unknown subjective test databases are not available, more sophisticated methods such as K-fold and leave-one-out cross-validation or bootstrapping are used (Witten and Frank 2005).

In the following, *training* and *test* sets refer to the datasets used for training and evaluating the model performance respectively. The metrics proposed here for computing the model performances are more thoroughly described in ITU (2012i).

Metrics are usually computed based on quality scores per sequence. All metrics detailed below may also be applied using the scores averaged per condition. This is for instance useful when the impact of the content has to be excluded from the analysis.

2.4.3.1 Variance Between Subjective Experiments

Due to the biases inherent to the subjective tests, quality scores may be shifted or distorted from one test to another one, especially if no anchor conditions have been used. The models should not mimic this inter-test variation. As a consequence, it is common practice to map the predicted scores to the subjective scores by applying a polynomial mapping. In the P.NAMS and P.NBAMS campaigns, a first-order polynomial mapping was applied, using Eq. (2.4):

$$y = a \cdot x + b \tag{2.4}$$

where y represents the subjective ratings averaged per sequence, x the outputs of the model for the corresponding sequence, and a and b the curve-fitting coefficients. In other studies, as for instance in VQEG or during the PESQ (ITU 2011b) or P.OLQA (ITU 2011b) development campaigns, a third-order polynomial mapping was used, as shown in Eq. (2.5):

$$y = a \cdot x^3 + b \cdot x^2 + c \cdot x + d \tag{2.5}$$

As for Eq. (2.4), y represents the target values and the subjective ratings averaged per sequence, x the outputs of the model for the corresponding sequence, and a, b, c and d the curve-fitting coefficients.

The third-order polynomial mapping compensates for the saturation effect that sometimes occurs with bounded scales such as the 5-point ACR quality scale (see Sect. 2.3.1).

In the performance metrics described below, this mapping is impacting the degrees of freedom, i.e., the number of values that are free to vary in the computation. In the case of a first-order mapping, the degrees of freedom (of the mapping function) are $d = 2$, and $d = 4$ for a third-order mapping.

2.4.3.2 Statistical Evaluation Metrics

Three criteria are commonly used for characterizing the model performance: its accuracy, consistency, and linearity/monotonicity (Wu and Rao 2006b). The accuracy captures the ability of the model to predict the subjective quality ratings with low errors. The consistency reflects the degree to which the model maintains prediction accuracy over the range of test sequences. Finally, the monotonicity corresponds to the degree to which the model's predictions agree with the relative magnitude of subjective quality ratings.

When the data is drawn from test data with near-normal distribution, these criteria are obtained by computing the prediction error, the outlier ratio, and the Pearson correlation coefficient respectively (ITU 2012i). When it cannot be verified that data are drawn from a near-normal distribution, the Spearman Rank coefficient is used in the literature instead of the Pearson correlation coefficient as a measure of monotonicity [see also Coolican (2009)].

The following metrics are computed per experiment, i.e. per test database. Then, for obtaining the model performance over all experiments, the output of each metric per database is averaged over all databases. For that purpose, different weights may be attributed to the databases. For instance, in the multimedia case, more weights may be attributed to the audiovisual databases than to the audio and video databases.

2.4.3.3 Model Accuracy

The prediction error (i.e., accuracy) is obtained using the Root Mean Square Error (RMSE):

$$rmse = \sqrt{\frac{1}{N-d}\sum_{i=1}^{N} Perror(i)^2}, \tag{2.6}$$

where N is the number of sequences used for comparing the estimated quality scores to the subjective scores, d is the degree of freedom of the mapping function (see above), and:

$$Perror(i) = Q(i) - Q_e(i). \tag{2.7}$$

here, i is the index of the sequence, Q is the subjective quality, and Q_e is the estimated quality.

This metric depends on the rating scale used during the subjective tests. As a consequence, for comparing two *rmse* values, the quality scores first need to be converted to the same scale. *rmse* is always positive, and lower *rmse* values indicate higher accuracy.

For taking into account the degree of uncertainty of the subjects' judgments, the so-called epsilon-modified Root-Mean-Square-Error ($rmse^*$) between the predicted and the subjective quality values can be computed instead of the *rmse*.

$rmse^*$ is similar to *rmse*, but with

$$Perror(i) = max(0, |Q(i) - Q_e(i)| - ci_{95}(i)), \tag{2.8}$$

where ci_{95} is the 95 % confidence interval of the sequence i. This metric is especially useful for comparing model performances based on subjective test databases with highly-varying confidence intervals, as was the case for the new ITU-T standard for full-reference speech quality assessment P.OLQA [Objective Listening Quality Assessment, ITU-T Recommendation P.863 (ITU 2011b)].

2.4.3.4 Model Consistency

The consistency of the model is obtained by computing either the outlier ratio or the residual error distribution.

The outliers are defined as the points for which the prediction error exceeds the 95 % confidence interval, i.e. if

$$|Perror(i)| > \frac{z \cdot \sigma(Q(i))}{\sqrt{N_s(i)}} \tag{2.9}$$

where $\sigma(Q(i))$ is the standard deviation of the scores associated with the sequence i, and $N_s(i)$ is the number of scores obtained for i. If $N_s(i) > 30$, then the Gaussian distribution can be used, and therefore $z = 1.96$. If $N_s(i) < 30$, the t-Student distribution is used instead, with $z = t$ and t the respective t-Student distribution variable, which depends on $N_s(i)$.

The outlier ratio OR is defined as the total number of outliers, N_{Out}, in relation to N, the total amount of data points (Eq. 2.10):

$$OR = \frac{N_{Out}}{N}. \tag{2.10}$$

2.4.3.5 Model Monotonicity

The Pearson Correlation Coefficient is used as a measure of the monotonicity of the model (i.e., linearity) and is expressed as follows:

$$R = \frac{\sum_{i=1}^{N}(Q(i) - \overline{Q}) \cdot (Q_e(i) - \overline{Q_e})}{\sqrt{\sum_{i=1}^{N}(Q(i) - \overline{Q})^2} \cdot \sqrt{\sum_{i=1}^{N}(Q_e(i) - \overline{Q_e})^2}} \tag{2.11}$$

This coefficient measures the linearity between the estimated and subjective quality scores.

The Spearman rank-order correlation coefficient is defined as follows:

$$R_S = \frac{\sum_{i=1}^{N}(RO(i) - \overline{RO}) \cdot (RO_e(i) - \overline{RO_e})}{\sqrt{\sum_{i=1}^{N}(RO(i) - \overline{RO})^2} \cdot \sqrt{\sum_{i=1}^{N}(RO_e(i) - \overline{RO_e})^2}} \tag{2.12}$$

This formula is similar to the Pearson Correlation Coefficient, apart from the fact that the rank order of the subjective ($RO(i)$) and estimated ($RO_e(i)$) quality scores are taken instead of the quality scores themselves. This metric thus measures if the increase (resp. decrease) of one variable is associated with the increase (resp. decrease) of the other variable, independently of the magnitude of the increase (resp. decrease). This metric is a non-parametric measure of monotonicity.

2.4.3.6 Statistical Significance Evaluation

A number of performance metrics have been defined above. One or several of these metrics can be used for comparing the prediction performance of different models. It is often required to compare the performance of different models, e.g., when selecting from different variants of a given algorithm, or to decide on the best algorithm in a standardization context. In this case, it is required that the better performance of the "winning model" is grounded on solid statistical analysis. Here, it is to be assessed whether the performance of the models is significantly different, e.g., undertaking the following steps:

To determine whether a performance index such as the **RMSE** is statistically significantly different for two models, the q statistic defined by Eq. (2.13) is first computed (ITU 2012i):

$$q = \frac{rmse_h^2}{rmse_l^2} \tag{2.13}$$

here, $rmse_h$ and $rmse_l$ are the highest and lowest $rmse$ values considered in the comparison. The q statistic has an F-distribution, with n_1 and n_2 degrees of freedom. n_1 and n_2 are given by $N_1 - d$ and $N_2 - d$, where N_1 and N_2 are representing the total number of samples for the compared average prediction errors, and d is the degree of freedom of the mapping function (see above). The q statistic is evaluated against the tabulated value $F(0.05, n_1, n_2)$ that ensures 95 % significance level: if $q > F(0.05, n_1, n_2)$, the difference is significant.

The significance of the difference between the **outlier ratios** is obtained by comparing the normalized statistic Z_N expressed in Eq. (2.14) [Fisher-z transform, see Spiegel (1998)] to the tabulated t value of the t-student distribution for the 95 % significance level for the two tailed test. If $Z_N > t$, then the compared outlier ratios $OR1$ and $OR2$ are statistically significantly different with 95 % significance level.

$$Z_N = \frac{OR1 - OR2 - \mu_{OR1-OR2}}{\sigma_{OR1-OR2}} \tag{2.14}$$

here, μ is the mean and σ the standard deviation, i.e.
$\mu_{OR1-OR2} = \mu(OR1) - \mu(OR2) = OR1 - OR2$,
and

$$\sigma_{OR1-OR2} = \sqrt{\frac{\sigma(OR1)^2}{N_1} + \frac{\sigma(OR2)^2}{N_2}} \tag{2.15}$$

Under the $H0$ hypothesis that $OR1$ and $OR2$ are equal ($OR1 = OR2 = OR$), the mean becomes

$$\mu_{OR1-OR2} = 0, \tag{2.16}$$

and the standard deviation becomes

$$\sigma_{OR1-OR2} = \sqrt{OR \cdot (1 - OR) \cdot \left(\frac{1}{N_1} + \frac{1}{N_2}\right)} \tag{2.17}$$

Similarly, the significance of the difference between two **Pearson correlation coefficients** is obtained by comparing the Z_N statistics to the two-tailed t value

$$Z_N = \frac{z_1 - z_2 - \mu_{z_1 - z_2}}{\sigma_{z_1 - z_2}} \tag{2.18}$$

z_1 and z_2 are the Fisher-z transformed values of the compared Pearson correlation coefficients. Under the $H0$ hypothesis that z_1 and z_2 are equal, the mean $\mu_{z_1-z_2} = 0$, and the standard deviation $\sigma_{z_1-z_2} = \sqrt{\sigma_{z_1}^2 + \sigma_{z_2}^2}$. The standard deviation of the z-fisher statistics is shown in Eq. (2.19):

$$\sigma_z = \sqrt{\frac{1}{N-3}} \tag{2.19}$$

If Z_N is below the two-tailed t value, the hypothesis $H0$ is true and the two correlation coefficients are not significantly different. If $Z_N > t$, the coefficients are significantly different.

For evaluating the performance of a quality model, the above metrics should be applied on unknown test sets, i.e., on data which have not been used for training the model. The test sets should be representative of the scope the model is supposed to cover. For instance, the test sets used for evaluating the performance of a video quality model in the context of IPTV should cover degradations, video contents, formats and codecs typical of IPTV services. However, oftentimes the model developer has usually access to a limited amount of data so that it becomes problematic to split this limited amount of data into a training and a test set. One common way of circumventing this issue is cross-validation (Witten and Frank 2005). Cross-validation consists in splitting the available data into several (k) folds. $k - 1$ folds are used for training the model. The resulting model is tested on the remaining fold. This procedure is repeated k times. The split into folds can be done randomly, but it is recommended to *stratify* the folds. In the context of quality testing, this means for instance that each fold should cover typical degradations and quality range for the targeted application. Indeed, if a model is for instance trained only on compression artifacts, it cannot predict well the quality when there is packet-loss. For making the procedure more robust, in particular to make it independent of the selected k folds, the whole procedure is typically run several times. Common practice is to use $k = 10$, and to run the stratified 10-fold cross-validation 10 times, resulting in 100 runs. The performance metrics are computed for each run, and the average over these runs are taken for each metric.

One form of the cross-validation is the leave-one out cross-validation, where $k = n$, and n is the number of samples in the dataset. However, due to the similarity between the training sets, these performance metrics will probably have too high variance (Hastie et al. 2009).

An alternative to the cross-validation is the bootstrap method. One variant of this method consists in sampling a dataset of n elements n times with replacement, resulting in another dataset of n elements. Since the sampling is done with replacement, a given element can be sampled several times, while other elements will not be picked from the original dataset. As a consequence, some elements will appear multiple times while others will not be in the new dataset. The non-selected datasets are used for testing the model performance. It can be proven that statistically and for a reasonable dataset size, 63.2 % of the elements of the original dataset will be in the new dataset (the training set). The remaining elements (36.8 % of the original dataset) will be in the test dataset. Performance metrics can then be computed on the test set ($perf_{test}$). However, due to the relatively small size of the training dataset (63.2 % of the original dataset), the performance results may be too pessimistic. As a consequence, these performance metrics are also computed on the training dataset ($perf_{training}$), and a weighted average ($perf_w$) is taken, as shown in Eq. (2.20).

$$perf_w = 0.632 \cdot perf_{test} + 0.368 \cdot perf_{training} \tag{2.20}$$

This procedure is repeated several times, each time with different replacement statements, and the average of the performance results obtained with Eq. (2.20) is computed.

2.5 Summary

This chapter presents all background information required for developing an audio-visual quality model for IPTV services, referred to in this work as the "T-V-Model", following the proposed quality engineering flow chart shown in Fig. 2.1. Since the model has to ultimately estimate quality, definitions of quality and quality-related concepts such as QoS, QoE, quality elements, features and dimensions have been provided.

The T-V-Model should be able to estimate the perceived quality for typical IPTV degradations. For that purpose, the whole end-to-end transmission chain has been described, and degradations that each processing step may generate have been identified, as well as the perceptual effect and quality impact they may cause.

Since the model is developed to estimate the quality as perceived by the user, the T-V-Model will be trained on subjective quality results, quality ratings being considered as the "ground truth" for the perceived quality. However, many sources of biases may be present in these tests. Based on a literature review, potential biases related to standardized test methods and rating scales have been summarized.

Finally, based on a literature review, an overview of different model types has been provided, as well as modeling techniques and model evaluation metrics. Along with this overview, a framework for model development has been proposed.

Chapter 3
Model Framework

This chapter re-addresses the different steps of the quality engineering flow chart (Fig. 2.1), but focuses this time on the one-way IPTV services and details the choices made at each stage of the modeling development. In particular, after having defined the scope and type of the model(s), the quality elements and quality features relevant for the targeted application will be identified. The choice of subjective test methods and rating scales will be described and the selection of audiovisual contents, test design, and processing chain used for the tests will be detailed. Finally, the modeling method and the evaluation procedure followed for developing and validating the quality models are summarized.

Three sets of databases are described: the TVM, the P.NAMS training, and the P.NAMS validation database sets. The TVM set was entirely developed within the Deutsche Telekom Laboratories (also called "T-Labs" or "DT" in the following) during the T-V-Model project. The two P.NAMS database sets were produced by the ITU-T P.NAMS and P.NBAMS participants for the higher resolution application area (IPTV, VoD) within the ITU-T P.NAMS and P.NBAMS competitions. These have lead to the ITU-T standards P.1201, P.1201.1, P.1201.2, P.1202, P.1202.1, and P.1202.2 (ITU-T Recommendation P.1201 (2012); ITU-T Recommendation P.1201.1 (2012); ITU-T Recommendation P.1201.2 (2012); ITU-T Recommendation P.1202 (2012); ITU-T Recommendation P.1202.1 (2012); ITU-T Recommendation P.1202.2 (2012)). The model standardized in P.1201.2 is the one developed in this book. The audio model slightly differs but it performs as well as the audio module of the P.1201.2 standard. The three data sets have the same scope, and they were designed for developing and validating the same types of model. More details on these sets are described later in this chapter.

M.-N. Garcia, *Parametric Packet-Based Audiovisual Quality Model for IPTV Services*, T-Labs Series in Telecommunication Services, DOI: 10.1007/978-3-319-04855-0_3,

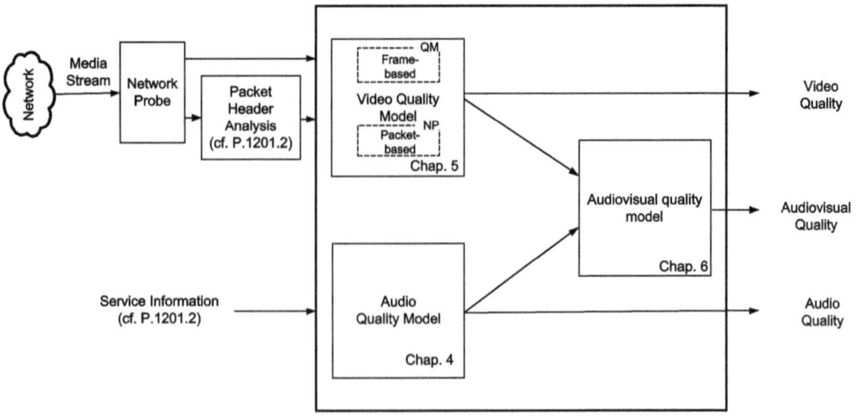

Fig. 3.1 Overview of the audiovisual quality model, indicating the sections of the book that describe the respective model algorithms. *NP* Network Planning; *QM* Quality Monitoring

3.1 Scope and Model Type

The scope of this work is to develop an audiovisual quality model for IPTV services. The model is to be used both for network planning and quality monitoring in the case of encrypted data streams. The model structure is shown in Fig. 3.1.

In the case of network planning—i.e., where the service is not yet set up—, information (input parameters) sent to the model are assumptions and choices made by the network planner: codec type and profile, expected bitrate, packet-loss-rate, packet-loss-distribution, and packet-loss-concealment. As input the model takes a low-level parametric description of the end-to-end audio and video processing and transmission path. The model at this stage is helpful for selecting the most appropriate codec parametrisation in terms of audio and video compression choices for the targeted application. It is also useful for estimating which amount of packet losses can be supported by the end-to-end network before becoming perceivable by the user, and can help to choose the best suited error resilience approach and settings.

In the case of quality monitoring, the service is set-up, and measurements are to be done on the running network. Here, more parameters can be extracted than in the case of network planning. The model can be implemented at different locations in the network. However, for having a better estimation of the quality experienced by the user, the model, or at least the input parameter extraction probe, should preferably be placed at the receiver side. This allows the model to consider all degradations occurring along the whole transmission chain. In the following, the realistic case of not having access to the source (undegraded) signals and of encrypted data is considered. The model does not have access to the payload, and therefore only relies on packet header information. Typical inputs are bitrate and packet-loss-rate, as for the network planning case, but here these inputs are derived from network measurements. Some additional information such as the frame type and frame size can be estimated, allowing the computation of more accurate parameters such as the spatial extent and the duration of loss-related video degradation.

As shown in Fig. 3.1, the audiovisual quality model, which is described in Chap. 6, is constituted of an audio quality model and a video quality model. Both models (audio and video) take a parametric description of the processing and transmission path as input, from the encoder to the decoder. The audio and video quality models are described respectively in Chaps. 4 and 5. The same version of the audio quality model is used for both network planning and quality monitoring. In contrast, different versions of the video quality model are used for the two application types. The version used for network planning also served as baseline during the development of the quality monitoring model. Since the models were ultimately trained based on the P.NAMS subjective test results, with a stimulus duration of 10 s, a measurement window of 10 s is used for extracting the model input parameters.

3.2 From Quality Elements to Quality

Quality elements and the associated parameters and quality features selected for the IPTV case are presented in this section. The parameters are, at this stage, not the input parameters of the model, but the parameters to be used for simulating different types and levels of degradations. The resulting impaired audio, video, and audiovisual sequences were used in the subjective audio, video, and audiovisual quality tests. The audio, video, and audiovisual quality models have been developed and evaluated based on the results of these tests. Identifying the most relevant quality elements and features—both in terms of quality and in the light of the targeted IPTV application—is thus a primordial step. This is the focus of this section. Quality elements considered in the modeling task will be summarized per data set (TVM, P.NAMS training and validation) in Sect. 3.6, Table 3.25. For more detailed information on the quality elements and features, the reader is referred to Sect. 2.2.1.

At the source level, the audiovisual signal is considered to be un-degraded. The recording quality is assumed to be perfect, and the pre-processing of the audiovisual signals such as color space conversion for video and sampling and quantization for audio are considered to be transparent. The source quality is thus assumed to be excellent, and the impact of the source quality on the perceived quality at the user side is not considered in the modeling.

In the case of IPTV services and high resolution videos (SD: 720 × 576 pixels (PAL) or 720 × 480 pixels (NTSC), HD720: 1280 × 720 pixels and HD1080: 1920 × 1080 pixels), the frame rate varies between 24 and 60 fps. These frame rates are expected to be sufficiently high for not degrading the perceived quality. However, for ensuring that the model covers this range of frame rates, contents of different frame rate will be used in the subjective quality tests. Also, since 24/25 fps contents may appear less fluid than 50/60 fps contents, different subjective tests are conducted for 24/25, 30, and 50/60 fps contents. Similarly, both interlaced and progressive contents will be processed for the tests, so that the developed model covers both scanning modes. The quality impact due to the video contents will be considered by using contents with various spatio-temporal activities and representative of different TV programs (see Sects. 3.3.2.1 and 3.3.3.1). Finally, the impact of the picture size (PAL, NTSC, HD720, HD1080) will be taken into account by conducting separate tests for the different formats. Note that all contents are shown to the subjects in their native resolution. No up-scaling is performed. Up-scaling and its impact on quality is treated separately in Garcia and Raake (2010).

The audio source signal is characterized by its sampling rate (44.1, 48 kHz), the number of recorded audio channels (mono, stereo), and the content type, i.e., if the audio contains speech, or speech on noise, or music, etc. The quality impact due to the number of recorded channels and its consideration in the audio quality model is let for further studies. Both mono and stereo signals were allowed in the conducted subjective tests, but they were presented in a stereo mode. The effect of the audio content type is covered by using various audio contents in the subjective tests (see Sects. 3.3.2.1 and 3.3.3.1).

At the encoding stage, the main elements affecting quality are the employed codec type and profile, the GOP structure, the slice size and the media bitrates (see below). Other quality elements such as the rate control, which heavily depends on the codec implementation, are not accessible for inclusion into the model.

In the case of IPTV, H.264 is mainly used for encoding both HD and SD video. MPEG-2 is also employed in the case of satellite transmission for SD video. Constante BitRate (CBR) seems also to be commonly used, and is the case considered in this work. Other elements impacting quality—and studied in this work—are the number of slices per frame used for encoding the video, and the GOP structure, which includes the GOP length, i.e., the number of I-frames per second, as well as aspects such as the number of B-frames between P-frames, or the use of reference B-frames as with hierarchical coding.

Several audio codecs are used in IPTV and will be considered in this work: MP2, MP3, AAC-LC, HE-AACv2, and AC-3.

Once the video and audio are encoded, they are multiplexed into transport streams, in our case MPEG2-TS, then into RTP, UDP, and IP. No error resilience algorithms such as FEC have been used. At the transmission level, relevant information is the packet loss amount and pattern. Delay is either compensated by the jitter-buffer or results in packet loss. Similarly, any error correction or packet loss handling algorithms such as retransmission mechanisms [e.g., Automatic Repeat reQuest (ARQ)] that may have an impact on the packet loss amount and pattern are not considered here. As a consequence, in this work, parameters related to the packet-loss amount and pattern will be measured at the input of the decoder, i.e., after all possible error or packet loss handling has been applied. For further information on the mapping between packet-loss-rate and packet-loss-distribution before and after packet-loss-handling mechanisms such as FEC and Automatic Repeat reQuest (ARQ), see the activity of ITU-T SG12 Q.17 on the performance parameters for stream repair techniques (ITU-T Recommendation Y.1540, Appendix VII).

At the video decoder level, the quality impact due to packet loss widely depends on the selected Packet-Loss Concealment (PLC) (slicing, freezing), which is a key

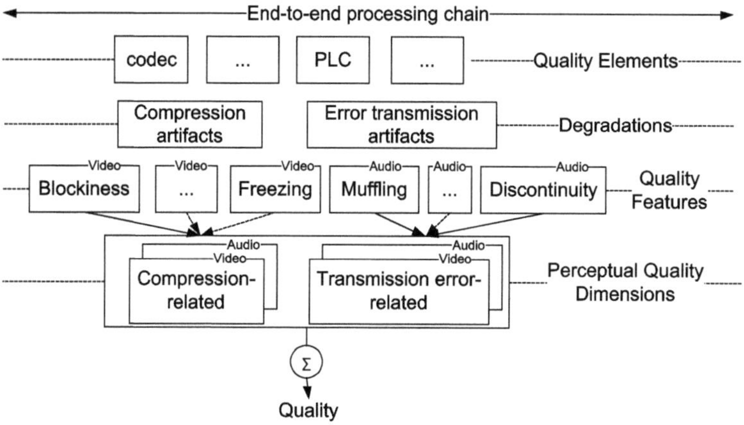

Fig. 3.2 From quality elements to quality in the case of IPTV

quality element. The impact of packet loss in the case of slicing further depends on the slicing configuration, i.e., on the number of slices per frame. Indeed, since the decoder re-synchronizes its decoding at the next available slice, the spatial extent of the loss increases with the slice size. In both the slicing and the freezing case, the loss duration depends on the GOP size, since losses occurring on I- and P-frames are propagating till the next I-frame, i.e., till the next GOP starts, that may be due to the next scene cut, if it is encoded with an I-frame.

The audio packet-loss-concealment approaches considered in this work are silence insertion and frame repetition. It is assumed that they are the most commonly used ones in IPTV services.

Additional degradations potentially introduced by post-processing elements such as de-interlacing algorithms are ignored. As previously mentioned, video up-scaling is not considered, nor are any degradation possibly introduced by the display. LCD displays were used for video subjective tests and audio was rendered in stereo on headphones or loudspeakers, depending on the tests.

The impact of audio and video quality on the overall audiovisual quality was studied in different audiovisual quality tests. To this aim, different levels of audio and video qualities for various audio and video degradation types and for various audiovisual contents have been used.

As shown in Fig. 3.2, the dominant degradations resulting from the end-to-end transmission system and covered in this work are compression and packet loss. As studied in Chap. 2, Sect. 2.2, the corresponding quality features in the case of video are blockiness/blurriness, slicing and freezing, the latter two depending on the applied packet-loss-concealment. It is assumed that these quality features translate into one compression-related quality dimension and one of two transmission-error-related dimensions, slicing-error or freezing-error. The perceived quality is a combination of these two dimensions. Note that in the TVM and P.NAMS tests, the decoder was not switching between slicing and freezing for a given impaired video. Hence, instead of using two separate transmission-error-related dimensions in addition to the compression-related dimension, one for slicing and one for freezing, one transmission-error-related impairment has been used.

The audio quality features corresponding to audio compression and packet-loss are birdie, muffled audio, audio discontinuity, and audio frame repetition. Similarly to video, it is assumed that these quality features translate into one compression-related dimension and one transmission-error-related dimension, and that the perceived quality is a combination of the two dimensions. As for video, the audio decoder was not switching between frame repetition and silence insertion for a given impaired audio file. One transmission-error related dimension was thus sufficient for capturing the quality impact due to packet loss for a given audio file.

Analytical tests and multidimensional analysis have not been conducted during the present work. As a consequence, the selected quality features and the perceptual quality dimensions were not empirically tested, i.e., via means of analytical tests and multidimensional analysis, but rather derived from the observed typical audio and video degradations and a respective literature review. The validity of the approach will indirectly be tested by evaluating the performance of the quality models developed based on the assumed perceptual quality dimensions.

3.3 Subjective Quality Assessment

The perceptual quality dimensions, or at least the quality features the models should cover, have been defined above in term of impairment terms. Since the models are developed based on the results of subjective quality tests, these quality tests should well cover the identified quality features. They should also be designed for avoiding any of the biases detailed in Chap. 2, Sect. 2.3. The present section gives an overview of all subjective tests conducted for developing and validating the audio, video, and audiovisual quality models. The used audiovisual contents and test design are described, and information is given about the processing chain used for generating the stimuli to be rated by the subjects. Finally, the test set-up is described, and details on the selected test method and rating scale are provided.

3.3.1 Database Overview

As mentioned in the introduction of this chapter, three sets of databases were used for developing the audio, video, and audiovisual quality models: the TVM and the P.NAMS training and validation database sets. The entire database set for P.NAMS was produced under NDA between the participants of the P.NAMS and P.NBAMS competitions.

Table 3.1 TVM database set

#	ID	Modality
1	$ph2_A$	Audio
2	$ph3_A$	Audio
1	ref-HD	Video (HD)
2	ref-SD	Video (SD)
3	test1-HD	Video (HD)
4	test1-SD	Video (SD)
5	test2-HD	Video (HD)
6	test2-SD	Video (SD)
7	test3-HD	Video (HD)
8	test3-SD	Video (SD)
9	test4-HD	Video (HD)
10	test4-SD	Video (SD)
11	test5-HD	Video (HD)
12	test5-SD	Video (SD)
1	$ph2_{AV}^{HD}$	Audiovisual (HD)
2	$ph2_{AV}^{SD}$	Audiovisual (SD)
3	$ph3_{AV}^{HD}$	Audiovisual (HD)
4	$ph3_{AV}^{SD}$	Audiovisual (SD)

HD format 1080p24 and 1080i25
SD format 576p24 and 576i25

The **TVM database set** is composed of two audio-only, twelve video-only (six SD, six HD) and four audiovisual subjective tests (two SD, two HD). These tests are listed in Table 3.1.

In the **P.NAMS training tests**, one audio-only, five video-only, and three audiovisual quality tests were conducted by the P.NAMS and P.NBAMS proponents. One additional audio test ($tr16$ in Table 3.2) was conducted at T-Labs for further developing the audio quality model. The resulting databases are listed in Table 3.2 along with the modality (audio, video, audiovisual) they address and the producer/owner of the databases. The following identifiers are used for designating the owners of the databases: YON: Yonsei University, South Corea; DT: Deutsche Telekom, Germany; NTT: Nippon Telegraph and Telecom, Japan; NS: Netscout, UK; TEC: Technicolor, China; and TEL: Telchemy, USA. This reflects the way in which the training databases have been created: For a given database of the P.NAMS training database set, the set of content selection, test plan generation, processing, and conducting of subjective test have been performed by a single proponent (indicated by the respective identifier).

In the **P.NAMS validation tests**, two audio-only, eight video-only, and four audiovisual quality tests were conducted by the P.NAMS and P.NBAMS proponents. The resulting databases are listed in Table 3.3 along with the modality (audio, video, audiovisual) they address, the producer/owner of the databases, and the proponent who processed the sequences. In contrast to the P.NAMS training tests, the test plan design and the processing of source files were not performed, for a given database, by the same proponent (column "Owner") as the source selection and the

Table 3.2 P.NAMS training database set

#	ID	Modality	Owner
1	tr07	Audio	YON
2	tr16	Audio	DT
1	tr08	Video (HD, 1080p30)	NTT
2	tr09	Video (SD, 576i25)	NS
3	tr10	Video (HD, 720p50)	DT
4	tr11	Video (HD, 1080i30)	YON
5	tr15	Video (HD, 1080p25)	TEC
1	tr12	Audiovisual (SD, 576i25)	NS
2	tr13	Audiovisual (HD, 720p50)	DT
3	tr14	Audiovisual (HD, 1080i30)	NTT

Table 3.3 P.NAMS validation database set

#	ID	Modality	Owner	Processing
1	vl11	Audio	NTT	NS
2	vl12	Audio	TEL	NTT
1	vl13	Video (SD, 576i25)	NS	TEL
2	vl14	Video (HD, 720p50)	DT	TEC
3	vl15	Video (HD, 1080p25)	TEC	DT
4	vl16	Video (HD, 1080i25)	TEL	YON
5	vl17	Video (SD, 480i30)	NTT	YON
6	vl18	Video (HD, 720p60)	YON	DT
7	vl19	Video (HD, 1080p30)	YON	NTT
8	vl20	Video (HD, 1080i30)	NS	TEC
1	vl21	Audiovisual (SD, 480i30)	NTT	NS
2	vl22	Audiovisual (HD, 720p60)	YON	DT
3	vl23	Audiovisual (HD, 1080p25)	DT	YON
4	vl24	Audiovisual (HD, 1080i25)	DT	NTT

conducting of the subjective test (column “Processing”). This procedure prevents a given proponent for having both the processed files to be sent to the models (here, Packet CAPture (PCAP) files) and the subjective quality scores corresponding to these files, and therefore for adapting the subjective scores to the model output.

Details on the audiovisual contents and conditions used in the tests are given in the following sections.

3.3.2 TVM Data Set

3.3.2.1 TVM Audiovisual Contents

The source material used in each database of the TVM set consists in five audiovisual contents of 16 s duration each. Such short stimulus durations are used to avoid the

Table 3.4 Audiovisual contents description of TVM databases

ID	Name	Video	Audio
1	A	Movie trailer, several scenes	Speech on music (German male speech over soft movie music)
2	B	Interview, one scene	German speech (interview), female
3	C	Soccer, one scene	Speech on babble noise from football crowd
4	D	Movie, one scene	Classical music (Harpsichord and strings)
5	E	Music video, several scenes	Pop music with singer

recency effect (see Sect. 2.3.1). Note that 5–10 s duration sequences are commonly used for the subjective tests underlying speech, audio, and video quality model development campaigns run by VQEG or ITU-T, as well as for speech, audio, and video coding standardization-related evaluation campaigns. However, a 16 s duration was found more appropriate for the assessment of HD video sequences. Indeed, due to the bigger size of HD sequences, subjects may require more time for appreciating and judging the whole video. For the sake of homogeneity and for being able to compare results for the two formats, the same duration was kept for SD videos.

Video-only and audiovisual tests were conducted separately for the two video resolutions SD and HD. The audiovisual contents are representative of different TV-programs. The video contents differ in their amount of detail and complexity of structures and movements, and the audio contents in terms of audio category and genre. The selected audiovisual content types are described in Table 3.4.

As indicators of the spatio-temporal complexity of the content, the spatial perceptual information (SI) and the temporal perceptual information (TI) as defined in the ITU-T Recommendation P.910, and using Eqs. (2.1) and (2.2) (see Sect. 2.2.1.2), are shown for the five HD TVM contents in Fig. 3.3. The temporal complexity of contents "A" and "D" is well captured by the TI values. These two contents indeed contain complex (chaotic) movements which are expected to be difficult to encode. Similarly, content "C" (soccer) has a complex structure (the grass of the soccer field), which seems to be well captured by the SI value.

The audio part of these contents was used for the audio-only quality tests. Similarly, only the video part was presented to the subjects in the video-only quality tests. Finally, the whole audiovisual contents were used in the audiovisual quality tests. In all tests, subjects were asked to rate the overall quality. This means, in particular, that subjects did not rate the audio-only and video-only quality in the audiovisual quality tests, but only the overall audiovisual quality.

3.3.2.2 TVM Test Design

Test conditions differ from test to test, but they all address typical degradations of IPTV services.

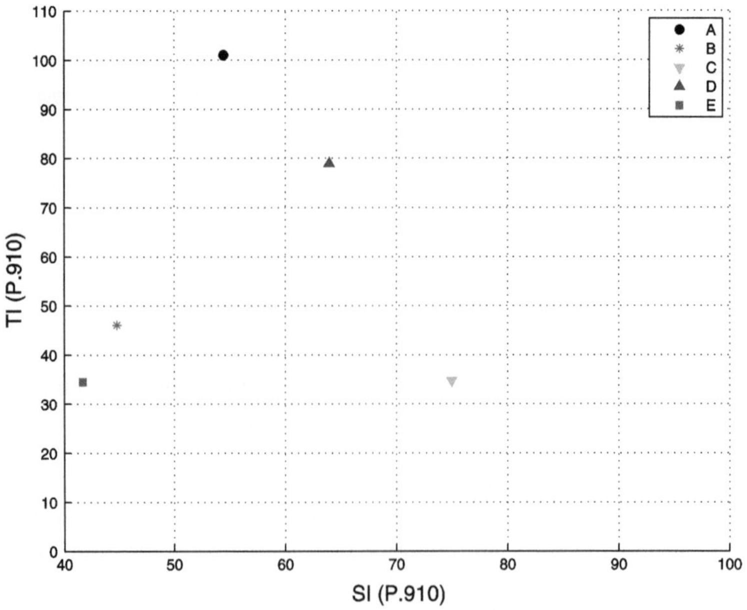

Fig. 3.3 Temporal perceptual information (*TI*) versus spatial perceptual information (*SI*), according to ITU-T Recommendation P.910 (1999) and using Eqs. (2.1) and (2.2). TVM contents (*HD*)

For video, those degradations are due to coding and compression, yielding blocking artifacts or blurriness depending on the de-blocking filter, or due to transmission channel errors, yielding slicing or freezing artifacts, depending on the applied Packet-Loss-Concealment (PLC).

The bitrate and packet-loss ranges used in the **twelve video subjective quality tests** are summarized in Table 3.5. The test conditions were applied offline on the five video contents described in Table 3.4. Detailed test plans are given in Appendix II.

The bitrate and the packet-loss percentage have been adjusted for achieving the same quality range between SD and HD. As a result, and due to their lower resolution, the SD videos are encoded with a bitrate four times smaller than for HD videos. In addition, the packet loss percentage for SD is slightly higher (0.06 %) than for HD (0.02 %). Indeed, with their smaller bitrates, SD videos are transmitted with fewer IP packets than HD videos. With short sequences, as is the case for TVM video sequences, and low bitrate (e.g., 0.5 or 1 Mbps), the amount of transmitted IP packets becomes too low for achieving a packet-loss percentage of 0.02 %.

It should be noted that for the purpose of modeling, bitrates lower than the bitrates in real IPTV application have been used. This allows the user of the model to know from which bitrate onwards the quality becomes critical. Of course, too low bitrates should not be used in order to avoid the corpus effect described in Sect. 2.3.2. Similarly, packet-loss-rates higher than the actual packet-loss-rates measured in the real IPTV application are introduced.

Table 3.5 Test conditions used in the **TVM video tests** for HD and SD

Parameters	Description
Format	{HD: 1080p/i; SD: 576p/i} @24 /25 fps
Codec	H.264 MP; MPEG-2 MP
CBR	HD, H.264: {2, 4, 8, 16, 32} Mbps
	HD, MPEG2: {4, 8, 16, 32, 64} Mbps
	SD, H.264: {0.5, 1, 2, 4, 8} Mbps
	SD, MPEG2: {1, 2, 4, 8, 16} Mbps
Ppl, freezing	HD: {0, 0.02, 0.06, 0.125, 0.25, 0.5, 1, 2} %
	SD: {0, 0.06, 0.125, 0.25, 0.5, 1, 2} %
Ppl, slicing, 1 slice per MB-row	{0, 0.125, 0.25, 0.5, 1, 2, 4} %
Ppl, slicing, 3-slices-per-frame	{0, 0.25, 0.5, 1, 2, 4} %
PLD	Random and bursty (μ :{4, 16, 64})
PLC	Freezing; slicing

MP Main Profile
CBR Constant BitRate
Ppl Packet-Loss percentage, *MB* Macro-block
PLD Packet Loss Distribution
μ burstiness, averaged number of consecutively lost packets
PLC Packet Loss Concealment
When values differ between SD and HD, the video format (HD or SD) is indicated. Detailed test plans can be found in Appendix II. Tests were conducted separately for HD and SD. Processing details are provided in section "TVM Processing chain"

Table 3.6 Number of conditions per degradation types for each test of the **TVM video data set**. HD resolution

Degradation	Ref	Test1	Test2	Test3	Test4	Test5
None	1	1	1	1	1	1
Block/blur	15	6	6	8	10	5
Slicing	5	18	18	31	24	16
Freezing	3	11	11	15	8	12
Total	24	36	36	55	43	34

In all tests, 3 error patterns were used per content and condition. Each conditions is characterized by the packet-loss percentage (*Ppl*), the bitrate and the packet-loss-concealment (PLC). This choice was made to assess different combinations of loss patterns and trace, and results in various loss distributions and thus perceptual impacts. Error-patterns were generated using a 2-state Markov model with a good state (packet found) and a bad state (packet lost). A zero-motion-concealment was used as error-concealment.

Table 3.6 summarizes the number of conditions used in each HD test per degradation type. In this table, "none" corresponds to the uncompressed non-degraded source files. "Block/blur" degradations correspond to potential compression artifacts obtained by reducing the bitrate (with H.264 or MPEG2). "slicing" and "freezing" degradations indicate when "slicing" or "freezing" are used.

Table 3.7 Anchor conditions used in all **TVM video tests**, HD and SD

ID	Description	Degradation	Quality
1	H.264; 8 Mbps	Block/blur	Medium
2	Uncompressed	None	High
3	H.264; 16 Mbps; 4 % loss (random); slicing	Slicing	Low
4	H.264; 2 Mbps	Block/blur	Low
5	20 % of frames skipped: 0.8 s/1.6 s/0.8 s	Freezing	Low
6	Medium blurring	Blurring	Medium

For SD, the bitrate is four times smaller. Processing details are provided in section "TVM Processing chain"

The reference test ("Ref") addresses MPEG-2 and H.264 encoding, packet loss leading to slicing and freezing, several levels of blurriness, several GOP lengths. Tests 1 and 2 address both encoding artifacts and random packet loss, in the case of MPEG-2 and mainly H.264 encoding. Tests 1 and 2 are actually identical in terms of contents and conditions. Test 2 was conducted for compensating for processing issues in Test 1. Test 3 focuses on the impact of the loss distribution in the case of H.264 encoding, assessing both random and bursty losses characterized by the number of packets lost in a row (i.e., in terms of a 2-state Markov model), and on the impact of the number of slices per frame. Two encoder settings were used for that test: three slices per frame and one slice per Macro-block (MB) row (i.e., 68 slices per frames (34 slices per field) for progressive (interlaced) HD, and 36 slices per frame (18 slices per field) for progressive (interlaced) SD. Test 4 extends Tests 1, 2, and 3 to further conditions of packet loss, other GOP lengths and encoding profiles, in the case of H.264 encoding. Test 5 is similar to Test 1 and 2 in terms of conditions for the H.264 codec. However, it differs in terms of the realization of the test conditions. In particular, the processing chain used for impairing the video sequences is closer to the one used in the P.NAMS and P.NBAMS competitions. Details on the processing chains are provided in Sect. 3.3.2.3.

Six anchor conditions covering the whole range of quality and degradation types were used in all tests (mainly compression, freezing and slicing). This allows merging results from the different tests by mapping the results of each test to the other tests using the anchors. More details on the mapping between anchors are given in Sect. 3.4.2. The six anchors are listed in Table 3.7 for HD. Anchors for SD are identical to the HD case apart from the bitrate, which is four times smaller.

For developing the audio quality model, **two audio-only** subjective quality tests were conducted ($ph2_A$ and $ph3_A$). They both address compression and transmission error artifacts. The test conditions applied (off-line) to the audio source sequences described in Table 3.4 are summarized in Table 3.8.

Four codecs were used in both tests (MP2, MP3, AAC-LC, HE-AACv2). In the first audio test ($ph2_A$), no more than one audio frame was lost in a row and the

Table 3.8 Test conditions used in the **TVM audio tests**

Parameters	
Format	48 kHz, 16 bit, stereo
Codec	AAC-LC, MP2, HEAACv2, MP3
CBR	AAC-LC:{24, 32, 48, 64, 96} kbps
	HEAACv2: {16, 24, 32, 48, 64} kbps
	MP2: {48, 64, 96, 128, 192} kbps
	MP3: {32, 48, 64, 96, 128} kbps
Ppl	{0, 1, 4, 8}% (frame loss ≡ 1 frame per packet)
PLD	Random and bursty ($\mu = \{1,2,4,6\}$)
PLC	Codec built-in

CBR Constant BitRate
Ppl Percentage packet-loss
PLD Packet Loss Distribution
PLC Packet Loss Concealment
Detailed test plans can be found in Appendix II. Processing details are provided in section "TVM Processing chain"

Table 3.9 Anchor conditions used in all **TVM audio tests**

ID	Description	Degradation	Quality
1	Reference	None	High
2	Band limitation	3.5 kHz	Low
3	Band limitation	7 kHz	Medium
4	Distortion MNRU	12 dB	Medium
5	Heavy coding	AAC-LC 48 kbps	Low
6	Interruption	MP2 192 kbps 8 % Pfl	Low

MNRU Modulated Noise Reference Unit (ITU-T Recommendation P.810 (1996))
Pfl Percentage frame-loss
Processing details are provided in section "TVM Processing chain"

same type of loss conditions was applied for all four codecs. In the second test ($ph3_A$), bursty frame losses were addressed, this time only using the AAC and MP2 codecs, and one, two, four, and six audio frames lost in a row. In the second test, the audio coding bitrate ranges were extended towards lower bitrates. The loss was inserted at the audio frame level. In order to study the influence of the loss location on the perceived quality, four error patterns were used per content and condition, one condition being the combination of a codec, bitrate, frame-loss-rate, and frame burst-length (i.e., the averaged number of frames lost in a row, indicated by μ in Table 3.8). Each error pattern was rated by one fourth of the subjects.

Similarly to video, a number of anchors have been used in all audio tests. They are given in Table 3.9. They cover typical degradations used in MUSHRA tests as anchors so that the tests can be compared to previous tests, as well as typical IPTV degradations covering the whole quality range used in the tests.

Four audiovisual tests were conducted within TVM for developing the audiovisual quality models: two SD audiovisual tests ($ph2_{AV}^{SD}$ and $ph3_{AV}^{SD}$), and two

Table 3.10 Test conditions used in the **TVM audiovisual tests**

Parameters	Video	Audio
Format	HD: 1920 × 1080 SD: 720 × 576	48 Hz, 16 bit, stereo
Codec	H.264	MP2; AAC-LC
CBR	{2, 4, 8, 16} Mbps (HD) {0.5, 1, 2, 4} Mbps (SD)	AAC-LC: 48 kbps MP2: {48, 96, 192} kbps
Ppl	{0, 0.02, 0.06, 0.25, 1} % (freezing, HD) {0, 0.06, 0.25, 1} % (freezing, SD) {0, 0.125, 0.5, 4} % (slicing)	{0, 1, 4, 8}% (frame loss ≡ 1 frame per packet)
PLD	Random	Random
PLC	Freezing; slicing	Codec built-in

Detailed test plans can be found in Appendix II. HD and SD tests were conducted separately. Processing details are provided in section "TVM Processing chain"

HD tests ($ph2_{AV}^{HD}$ and $ph3_{AV}^{HD}$). The audio stream of the audiovisual stimuli were taken from the audio-only tests ($ph2_A$ for $ph2_{AV}^{[H,S]D}$, $ph2_A$ and $ph3_A$ for $ph3_{AV}^{[H,S]D}$) and the video-only streams from the video-only tests ("test1-HD" and "test1-SD" for $ph2_{AV}^{HD}$ and $ph2_{AV}^{SD}$, and "test2-HD" and "test2-SD" for $ph3_{AV}^{HD}$ and $ph3_{AV}^{SD}$).

The audio-only, video-only, and audiovisual tests span the same audio- and video-quality ranges. This avoids introducing bias in the audiovisual tests, and ratings of the audio- and video-only tests can directly be used as ratings for the audio and video streams of the audiovisual stimuli. This gives the opportunity to develop the audiovisual quality model based on subjective audio and video quality scores instead of values estimated by audio and video quality models, thus avoiding modeling the noise introduced by the audio and video quality models.

The audiovisual quality tests were designed so that for a given degradation type (compression, transmission error), each level of audio quality (low, medium, high) is combined with each level of video quality (low, medium, high). Additional conditions addresses the combination of audio and video degradations, for instance the combination of video compression with lossy audio. These test conditions (49 in total per test) are summarized in Table 3.10. They were used for processing the five source contents described in Table 3.4.

Similarly to the audio and video quality tests, anchors were used in all audiovisual quality tests. They are listed in Table 3.11 and correspond to a sub-sampling of the combination of the audio and video anchors.

3.3.2.3 TVM Processing chain

The TVM processing chain is schematically shown in Fig. 3.4.

In the TVM databases set, video files were encoded with the H.264 codec (Main Profile) using x264, revision r609 [x26]. For MPEG-2 encoding in *Test*1 and *Test*2,

Table 3.11 Anchor conditions used in all **TVM audiovisual tests**

ID	Audio	Video	Audiovisual quality
1	No degradation	No degradation	High
2	No degradation	H.264, 16 Mbps, 4 % loss (random) slicing	Low
3	No degradation	20 % skipped frames + freezing	Low
4	No degradation	H.264, 2 Mbps	Medium-low
5	No degradation	H.264, 8 Mbps	Medium-high
6	Band limitation, 7 kHz	No degradation	Medium-high
7	Interruption, MP2 192 kbps 8 % frame loss percentage	H.264, 16 Mbps, 4 % loss (random) slicing	Low
8	Heavy coding, AAC-LC 48 kbps	No degradation	Medium
9	Interruption, MP2 192 kbps 8 % frame loss percentage	No degradation	Medium-low

Processing details are provided in section "TVM Processing chain"

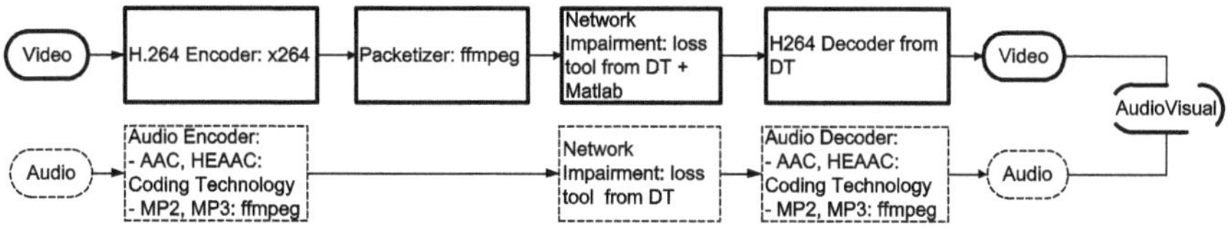

Fig. 3.4 TVM processing chain

ffmpeg was used instead. The encoded video stream was packetized into MPEG2-TS packets using a modified version of the open source ffmpeg tool (http://ffmpeg.mplayerhq.hu.). The TS packets were artificially grouped as if they were packetized into RTP/IP packets, with seven TS packets in each group. Packet loss impairments were applied on the whole TS groups by means of a proprietary tool. As mentioned earlier, for this dataset, no IP/UDP/RTP packet stack was added to the stream. As previously described, the video stream was impaired with both random and bursty losses. The pattern used for inserting bursty losses was a two-state Markov-Model. The resulting impaired video streams were decoded using a proprietary decoder which can handle packet loss in terms of either freezing or slicing and can decode sequences without crashing even for heavy packet loss (at least up to 4 %, for the example of HD).

For controlling the amount of blurring indicated for the respective *6th* anchor condition in Table 3.7, blurring was artificially introduced by using a box-blur filter in MEncoder (http://www.mplayerhq.hu.). In condition 5 (Table 3.7), freezing with skipping is artificially applied directly on the raw video sequences at three different temporal locations of the video. The first and third freezing occurrences last 0.8 s and the second 1.6 s. The resulting percentage of frozen and skipped frames is 20 %.

In the case of audio, ffmpeg (http://ffmpeg.mplayerhq.hu.) was used for encoding the audio files into MP2 and MP3. For AAC and HE-AAC, the AAC and HE-AAC encoder from Coding Technologies were used. At the time the audio tests were conducted, no tool was available for packetizing the encoded audio into MPEG2-TS and RTP packets. However, in the case of IPTV, it is likely that the loss of a packet containing audio yields the loss of a complete audio frame. As a consequence, with the use of a proprietary tool, complete audio frames were simulated to be lost by setting the data of the encoded frame to zero. The decoders detected such a frame as unusable and replaced them either by silence or by the previous available frame, depending on the decoder. With AAC and HE-AAC, it was possible to use frame repetition as loss concealment and was thus applied in most of the conditions. For MP2 and MP3, silence insertion was used by default. MP2 and MP3 audio streams were decoded with ffmpeg. For AAC and HE-AAC, the Coding Technologies decoder was used. This latter provides concealment that can be switched on and off (aacPlus v2 Evaluation Package v8.03). The concealment method is described in Ref. 3GPP TS.26.402.

Note that due to the lack of appropriate processing tools, video and audio were not multiplexed in the TVM databases, as is the case in IPTV. Since in IPTV the video bitrate is much higher than the audio bitrate, there are on average at least six video TSs in one IP packet. The TVM processing chain stays thus realistic. For audio, the TVM configuration is not as realistic as for video, but allows a better control of the achieved audio frame loss rate, which is useful for modeling purposes.

Further note that the TVM data were not encrypted. The encryption was simulated by using model input parameters which can be derived from PES- and TS-encrypted streams.

3.3.3 P.NAMS Data Set

3.3.3.1 P.NAMS Audiovisual Contents

There are two P.NAMS database sets: the P.NAMS training set and the P.NAMS validation set.

According to the ITU Q.14 P.NAMS and P.NBAMS test plans (ITU-T Recommendation P.1201/2 Supplement 2013), and as shown in Table 3.12, all video source sequences pertain to one of the following type: Movies/movie trailers, sports, music video, advertisement, animation, broadcasting news (business and current events), home video, and general TV material (e.g., documentary, sitcom, serial television shows).

The list below (see also Table 3.12) identifies the types of test material that form the basis for the selection of audio sequences: spoken words (e.g., read text (audio books), conversation, radio show, male and female voices), classical music (e.g., orchestras, piano, guitar), rock/pop music, advertisement (e.g., combination of music, sound effects, spoken text and singing), and dramatized plays (multiple speakers and background activity).

Table 3.12 Audio and Video **content types**, as specified by the ITU Q.14 P.NAMS and P.NBAMS test plan (ITU-T Recommendation P.1201/2 Supplement 2013), and used in the P.NAMS training and validation databases sets

Media	Content types
Video	Movies/movie trailers, sports, music video, advertisement, animation, broadcasting news (business and current events), home video, and general TV material (e.g., documentary, sitcom, serial television shows)
Audio	Spoken words (e.g., read text (audio books), conversation, radio show, male and female voices), classical music (e.g., orchestras, piano, guitar), rock/pop music, advertisement (e.g., combination of music, sound effects, spoken text and singing), and dramatized plays (multiple speakers and background activity)

Table 3.13 Content description, **P.NAMS training** database set

ID (Name)	DB	Video	Audio
1 (sto)	All	Documentary; 1 scene: fast panning of Stockholm taken from helicopter	Ambient music + ringing bells
2 (cro)	All	Cross-country; 3 scenes: flags, gun, cross-country	Music + crowd noise + gun shot
3 (wat)	All	Movie; 2 scenes; man in a fountain, duck on a lake	Ambient music + water fountain noise, music + duck sounds
4 (cla)	All	String quartet; 1 scene; left-right panning of a string quartet	Classical music (string quartet)
5-8 (n.a.)	All	n.a., content type as specified in Table 3.12	n.a., content type as specified in Table 3.12
5 (soc)	tr10, tr13, tr16	Sport; 1 scene: soccer game	Singing soccer fan + German reporter
6 (roc)	tr10, tr13, tr16	Pop music; 1 scene: rock show in a stadium	Rock/pop music
7 (tvn)	tr10, tr13, tr16	TV news; 1 scene: static camera	German TV news speaker
8 (ice)	tr10, tr13, tr16	Sport; 2 scenes: ice hockey fan, ice hockey game	Singing and clapping from ice hockey supporter

DB Database
n.a. non available
all all databases
tr13 audiovisual contents (*ID* 5–8) specific to the *tr*13 audiovisual database; the video stream of the *tr*13 contents is used for the "tr10" video-only database, and the audio stream for the "tr16" audio-only database

In the **P.NAMS training** set, eight 10 s duration contents (*SRCs*), as described in Table 3.13, were used per test.

Four of them are common to all databases—the common (CMN) set SRCs (*ID* 1–4 in Table 3.13). The remaining four contents (*ID* 5–8 in Table 3.13) are differing from one database to another one. They are provided in the table for the *tr*10 (video-only test), *tr*13 (audiovisual test), and *tr*16 (audio-only test) databases, which were produced at T-Labs. The contents *ID* 5–8 used in databases *tr*10, *tr*13, and *tr*16 are identical. However, only the video stream is used in *tr*10 while only the audio stream is kept for *tr*16. Both the audio and video streams are used in *tr*13. It was ultimately decided not to use content 3 (wat) in *tr*16 due to the "sound" made by the water fountain in the audio stream. Indeed, since only the audio was presented to the subjects in the audio-only *tr*16 test, the "sound" (or "noise") of the fountain could be mistaken for degradations by the subjects. Note that this issue was detected after having conducted the audio-only *tr*07 test. As a consequence, content 3 was part of the *tr*07 test, but quality ratings for this content were ignored in the data analysis and modeling. Three contents of the common set SRCs are extracted from the SVT ("Sveriges Television", the Swedish public service television company) multi-genre TV-program "Fairytale", mastered in progressive format at 3840 × 2160 pixels and 50 fps. These three contents were converted to the various formats of the P.NAMS databases. The fourth common set SRC was kindly provided under NDA by Orange Labs.

Special care was taken, mainly by visual inspection of the video sequences, for having the contents cover a wide range of spatio-temporal complexity. The "Spatial and Temporal perceptual Information" *SI* and *TI*, as defined in ITU-T P.910, was computed (see Sect. 2.2.1.2, Eqs. (2.1) and (2.2)). The *SI* and *TI* values for the *tr*10 database are shown in Fig. 3.5. Content "07-*tvn*" represents a TV-News speaker who barely moves, with a static background. This low complexity is well captured by the really low *SI* and *TI* values. The high *SI* and *TI* values of the "08-*ice*" and "02-*cro*" contents are validated by a visual inspection of the contents: They both contain a scene representing a crowd, and therefore many details, and complex movements, either when the crowd is moving in case of "02-*cro*", or from ice-hockey players moving on the ice field. The soccer content "05-*soc*" obtains similar values as the soccer content of the TVM database, with low temporal complexity but medium-high spatial complexity due to the complex structure of the grass. Contents "01-*sto*" and "04-*cla*" contain a smooth camera panning (of a town in content "01-*sto*", and of a string quartet in content "04-*cla*"). Since the panning is continuous and smooth, the movement is not considered as complex (in the computation of *TI*, it is the standard deviation of the movement which is considered, not its amplitude).

In the **P.NAMS validation** tests, eight 10 s duration contents were used per test, as listed in Table 3.14. Two of them (*ID* 2 and 4 in Table 3.14) are taken from the common set sources of the P.NAMS training set. Two other contents (*ID* 13 and 14) are common to all validation databases and were not used for the training database set. The remaining four contents (*ID* 15–18) differ from one database to another.

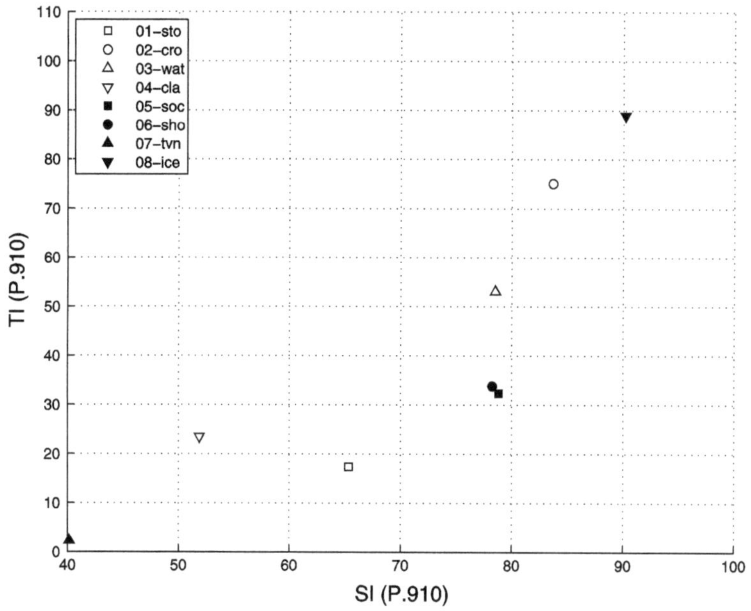

Fig. 3.5 Temporal perceptual information (*TI*) versus spatial perceptual information (*SI*), according to ITU-T Recommendation P.910 (1999) and using Eqs. (2.1) and (2.2). P.NAMS *tr*10 database

Table 3.14 Content description, **P.NAMS validation** database set

ID (Name)	Video	Audio
2 (cro)	Cross-country; 3 scenes: flags, gun, cross-country	Music + crowd noise + gun shot
4 (cla)	String quartet; 1 scene; left-right panning of a string quartet	Classical music
13 (ani)	Computer-generated movie; 5 scenes; girl running in a market place and meeting a creature	Movie-like music (string instruments)
14 (doc)	Documentary; 2 scenes; theater group running in a park; vertical, then horizontal panning	Ambient music + birdcall
15-18	As specified in Table 3.12	As specified in Table 3.12

3.3.3.2 P.NAMS Test Design

The conditions used in the P.NAMS databases follow the P.NAMS and P.NBAMS test plan (ITU-T Recommendation P.1201/2 Supplement 2013). This document regroups the specifications of the audio and video codecs and encoder settings (e.g., frame rate, GOP length), and of the range of bitrate and packet-loss rate to be used in the tests.

Table 3.15 Test conditions used in the **P.NAMS audio-only training** tests

Parameters	Values
Format	44.1 kHz, 48 kHz, 16 bit, mono and stereo, stereo headphone presentation
Codec	AAC-LC, MP2, HE-AACv2, AC3
CBR	AAC-LC: 32–576 kbps
	HE-AAC v2: 16–96 kbps
	MP2: 64–384 kbps
	AC3: 64–384 kbps
Ppl	0 % and [0.02, 5.5] % packet loss
PLD	Random and bursty (4-state-Markov Model)
PLC	Codec built-in

Table 3.16 Test conditions used in the **P.NAMS video-only training** tests

Parameters	Values
Format	HD: 1080p/i 25fps, 30fps, 720p50; SD: 576i25
Codec	H.264 (high profile for 1080, main profile for 720p and SD)
CBR	SD: 0.5–9 Mbps; HD: 0.5/1–30 Mbps
PLR	0 % and [0.011, 2] % (freezing)
	0 % and [0.011, 2] % (slicing, 1 slice/frame)
	0 % and [0.1, 2] % (slicing, 1 slice/MB-row)
PLD	Random and bursty (4-state-Markov Model)
PLC	Freezing and slicing

Tests were conducted separately for each video format

An overview of the conditions used in the P.NAMS training and validation tests is given in Tables 3.15, 3.16, 3.17, 3.18, 3.19 and 3.20.

Thirty conditions were used per P.NAMS test. For video and audiovisual tests, approximately 40 % of the conditions address compression-only artifacts, 40 % contain packet loss and apply slicing as PLC, and the remaining 20 % address packet-loss degradation with freezing as PLC. Both random and bursty packet loss were used in the tests, with a majority of bursty cases. Error patterns were generated using a four-state-Markov model (see section "P.NAMS Processing Chain" for more details). In the training database set, a single slicing configuration [either one slice per frame or one slice per Macro-Block-row (MB-row)] was used per test. In the validation tests, either a single slicing configuration (one slice per frame or one slice per MB-row) or both slicing configurations were used in a single test (but not in the same condition). For audio, approximately one third of the conditions address compression-only artifacts, and two thirds a combination of compression and packet loss degradations. Two to four audio codecs were used per audio or audiovisual database. Detailed test plans are given in the ITU.T P.1201 Recommendation.

It has already been mentioned that a common set SRCs were used for the P.NAMS database set. In addition, and similarly to the TVM set, common set conditions

Table 3.17 Test conditions used in the **P.NAMS audiovisual training** tests

Parameters	Values
Video format	SD: 576i25; HD: 720p50, 1080i30
Video codec	H.264 (high profile for 1080, main profile for 720p and SD)
Video CBR	SD: 0.5–9 Mbps
	HD: 1–30 Mbps
Audio format	44.1 kHz, 48 kHz, 16 bit, mono and stereo, stereo headphone presentation
Audio codec	AAC-LC, MP2, AC3
Audio CBR	AAC-LC: 32–576 kbps
	MP2: 64–384 kbps
	AC3: 32–96 kbps
Ppl	0 % and [0.5, 2] % (AAC-LC, slicing, 1 slice per MB-row)
	0 % and [0.017, 2] % (AAC-LC, freezing)
	0 % and [0.5, 2] % (AC3, slicing, 1 slice per MB-row)
	0 % and [0.017, 2] % (AAC-LC, slicing, 1 slice per frame)
	0 % and [0.017, 2] % (MP2, slicing, 1 slice per frame)
	0 % and [0.017, 0.25] % (MP2, freezing)
PLD	Random and bursty (4-state-Markov Model)
PLC	Freezing and slicing

Tests were conducted separately for each video format

Table 3.18 Test conditions used in the **P.NAMS audio-only validation** tests

Parameters	Values
Format	44.1 kHz, 48 kHz, 16 bit, mono and stereo, stereo headphone presentation
Codec	AAV-LC, MP2, HE-AACv2, AC3
CBR	AAC-LC: 48–576 kbps
	HE-AAC: 16–96 kbps
	MP2: 64–384 kbps
	AC3: 96–256 kbps
Ppl	0 % and [0.1, 6] % packet loss
PLD	Random and bursty (4-state-Markov Model)
PLC	Codec built-in

(HRCs) are also defined and applied to the common set SRCs for both the training and validation tests. The resulting processed files form the so-called "common set". The common set HRCs—or anchors—for video, audio, and audio-video are listed in Tables 3.21, 3.22, and 3.23. As for TVM, these conditions span the whole quality range and degradation types used in the tests.

Table 3.19 Test conditions used in the **P.NAMS video-only validation** tests

Parameters	Values
Format	1080p/i 25fps, 30fps, 720p50/60, 576i25, 480i30
Codec	H.264 (high profile for 1080, main profile for 720p and SD)
CBR	SD: 0.5–9 Mbps; HD: 0.5/1 to 30 Mbps
Ppl	Freezing: 0 % and [0.01, 0.8] %
	Slicing (1 slice/frame): 0 % and [0.01, 1] %
	Slicing (1 slice/MB-row): 0 % and [0.02, 1.2] %
PLD	Random and bursty (4-state-Markov Model)
PLC	Freezing and slicing

Tests were conducted separately for each video format

Table 3.20 Test conditions used in the **P.NAMS audiovisual validation** tests

Parameters	Values
Video format	480i30, 720p60, 1080p/i 25fps
Video codec	H.264 (high profile for 1080, main profile for 720p and SD)
Video CBR	SD: 0.5–9 Mbps; HD: 1 to 30 Mbps
Audio format	44.1 kHz, 48 kHz, 16 bit, mono and stereo, stereo headphone presentation
Audio codec	AAC-LC, MP2, HE-AACv2, AC3
Audio CBR	AAC-LC: 48 to 576 kbps
	HE-AACv2: 16 kbps
	MP2: 48–384 kbps
	AC3: 96–192 kbps
Ppl	0 % and [0.1, 2] % (AAC-LC, slicing, 1sl/MB-row)
	0 % and [0.01, 2] % (AAC-LC, freezing)
	0 % and [0.04, 1.01] % (MP2, slicing, 1sl/MB-row)
	0 % and [0.07, 1.03] % (MP2, freezing)
	0 % and [0.3, 0.6] % (AC-3, slicing, 1sl/MB-row)
	0 % and [0.125, 0.3] % (AC-3,freezing)
	0 % and 0.5 % (HE-AACv2, slicing, 1sl/MB-row)
	0 % and [0.02, 1.2] % (AAC-LC, slicing, 1sl/frame)
	0 % and [0.25, 1.03] % (MP2, slicing, 1sl/frame)
	0 % and 0.25 % (AC-3, slicing, 1sl/frame)
PLD	Random and bursty (4-state-Markov Model)
PLC	Freezing and slicing

Tests were conducted separately for each video format

3.3.3.3 P.NAMS Processing-Chain

The P.NAMS processing chain is shown in Fig. 3.6. More details can be found in ITU-T Recommendation P.1201/2 Supplement (2013).

As for TVM, the YUV (YCbCr) 4:2:0 video stream is H.264 encoded with the ×264 encoder revision r1867 [×26]. High profile was used as encoder setting for the HD1080 formats, and main profile for HD720 and SD. This H.264 encoded

Table 3.21 Anchor conditions used in all **P.NAMS video** tests

ID	Description	Degradation	Quality
1	H.264, very high BR, no loss	None	High
2	H.264, med. BR, no loss	Compression	Med-high
3	H.264, high BR, low bursty Ppl, slicing	Compression and slicing	Med
4	H.264, med BR, med burty Ppl, slicing	Compression and slicing	Med-low
5	H.264, high BR, high burty Ppl, freezing	Compression and freezing	Low

BR bitrate
Ppl Percentage packet-loss

Table 3.22 Anchor conditions used in all **P.NAMS audio** tests

ID	Description	Degradation	Quality
1	AAC-LC, 576 kbps, no loss	None	High
2	AAC-LC, med BR, no loss	Compression	Med-high
3	MP2, low BR, no loss	Compression	Med
4	AAC-LC, med BR, low bursty Ppl	Compression + interruption	Med-low
5	MP2, high BR, high bursty Ppl	Compression + interruption	Low

Table 3.23 Anchor conditions used in all **P.NAMS audiovisual** tests

ID	Description	Degradation	Quality
1	H.264 very high BR, AAC-LC, 576 kbps, no loss	None	High
2	H.264 med BR, MP2 low BR, no loss	Compression	Med-high
3	H.264 med BR, AAC med BR, low bursty Ppl, slicing	Compression + interruption + slicing	Med
4	H.264 high BR, AAC high BR, med bursty Ppl, slicing	Compression + interruption + slicing	Med-low
5	H.264 high BR, MP2 high BR, high bursty Ppl, freezing	Compression + interruption + freezing	Low

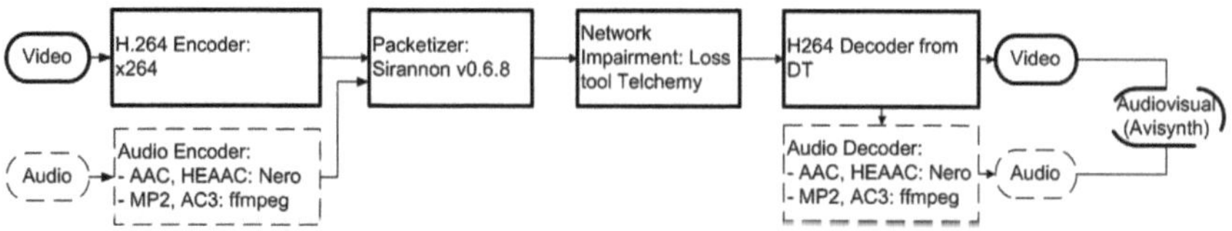

Fig. 3.6 P.NAMS processing chain

video stream is then packetized with the Sirannon v0.6.8 software (http://sirannon.atlantis.ugent.be.) into MPEG2-TS and RTP. To this aim, seven consecutive video TSs are grouped for forming one RTP/IP packet. The resulting Packet CAPture (PCAP) files are encrypted at TS- and PES-levels using a proprietary tool from Telchemy, USA. In the case of PES-level encryption, PES-headers are readable and only PES-payloads are encrypted. With TS-level encryption, MPEG2-TS headers are readable but MPEG2-TS payloads are encrypted, which includes the PES-headers and payload. The PCAP files are impaired with packet loss using a software from Telchemy. No losses are inserted in the first two seconds and the last second of the sequences. The loss pattern follows a four-state Markov-Model. The four states are "Random loss", "Error free", "Burst loss", and "Burst gap (no loss)". Some transition probabilities are pre-defined, and the remaining transition probabilities are derived from the targeted average loss probability, and the percentage of time in which the process resides in the gap state. This setting allows achieving both random loss and different strength of bursty losses with gaps (no loss) in the burst period. The impaired PCAP file is decoded with a proprietary H.264 decoder from T-Labs, which is actually an updated version of the decoder used in TVM.

In the case of audio, the signal level is first adjusted to −26 dBov. The resulting audio stream is encoded into MP2 or AC3 using the ffmpeg tool (http://ffmpeg.mplayerhq.hu). For encoding the audio into AAC-LC or HE-AACv2, the Nero encoder v1.5.4.0 (http://www.nero.com/eng/downloads-nerodigital-nero-aac-codec.php) was used. The encoded audio stream was multiplexed with a dummy video stream using the Sirannon software v0.6.8. The audio bitrate is six times lower than the video bitrate, so that one audio TS is, on average, packetized into one RTP packet. Encryption and packet loss insertion are realized using the same procedure as for video. The audio ES is extracted from the impaired PCAP file using the T-Labs decoder mentioned above. The audio ES is decoded with ffmpeg in the case of MP2 and AC3. For AAC and HE-AACv2, the audio ES is first packetized into a MP4 container with MP4Box (MP4Box-0.4.6-rev2735), and then decoded with the Nero AAC decoder.

In the case of AC3, frames are grouped by six into one "AC3-frame". As a consequence, when one audio frame is impaired with loss, the six audio frames of the "AC3-frame" are erroneous, yielding a burst loss of six frames. Since the audio bitrate is much lower than the video bitrate, it is not probable to loose more than six audio frames in a row. This motivated our choice of using six frames lost in a row as maximal burst length.

For audiovisual tests, the same video and audio processing chains as for the video and audio tests are used. The audio and video are streamed together, and the number of audio TSs and video TSs in a given RTP/IP packet depends on the audio and video bitrates. The impaired decoded audio and video streams are multiplexed into an AVI file using AVIsynth v2.5.

3.3.4 Test Set-Up

Listening and viewing conditions of the TVM and P.NAMS sets were compliant to ITU-T Recommendation P.800 (ITU-T Recommendation P.800 1996), and Recommendations ITU-R BT-500-11 (ITU-R Recommendation BT.500-12 2009) and ITU-T P.910 (ITU-T Recommendation P.910 1999) respectively. To ensure that the processed, then uncompressed material could be played out without playback artifacts, professional high-performance systems were used for audio and video presentation.

3.3.4.1 Listening Environment and Audio Calibration

In the TVM audio-only and audiovisual tests, two high-quality studio loudspeakers (Geithain RL 904) were used for presenting the audio, therefore reproducing a listening environment close to the IPTV usage scenario. The loudspeakers and participant ("subject") were located according to the ITU-R BS-1116-1 specifications for two-channel stereophonic systems (Fig. 6 in ITU-R BS-1116-1). The loudspeakers were therefore at the same height as the listener head. A 42″ LCD screen was placed in front of the listener between the loudspeakers. The screen was switched off in the case of the audio-only tests. The angle formed by the center of the screen, the listener and the right or left loudspeaker was 30° (see Fig. 3.7). For the P.NAMS audio and audiovisual databases, headphones were used to ensure similar listening environments across the different test labs. For the P.NAMS audio and audiovisual tests conducted at T-Labs, high quality headphones (AKG K271) were used along with a high quality sound card (Edirol UA-25). The quality impact due to the different listening environments between the TVM and P.NAMS audio tests could not be investigated due to lack of time, and it will be assumed in the following that this effect is negligible compared to the quality impact due to the audio degradations presented to the subjects.

Audio calibration in the TVM-audio-only tests were conducted as follows: According to ITU-R Rec. BS.1116-1, the calibration of the speaker system is done with a pink noise measurement signal with an *rms* equal to the "alignment signal level" of 18 dB below the clipping level. In the case of P.NAMS, the reference pink noise signal was adjusted to −26 dBov. This signal is fed into the input of each reproduction channel. For the TVM audio tests, the gain of the amplifier was adjusted to give a reference sound pressure level (IEC/A-weighted, slow) of −73 dBA for each of the loudspeakers, which results in a sound pressure of −76 dBA when both loudspeakers are switched on. For the audiovisual tests, the gain was adjusted to −62 dB instead of −73 dB for each loudspeaker, −73 dB being perceived as too loud for a TV-type viewing situation. In P.NAMS, the gain was adjusted to −73 dB for each ear of the headphone. When using loudspeakers in the tests, the sound pressure level was measured at the listener head's position with an NTI acoustic meter. When using headphones in the tests, a dummy head (Head Acoustics) is employed instead.

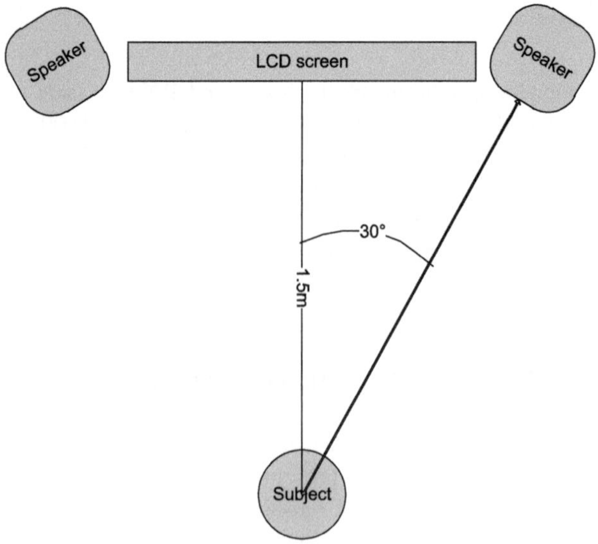

Fig. 3.7 Audio and audiovisual test set-up

Table 3.24 RMS level of TVM audio source contents

Item	Left (dB)	Right (dB)
A	−22.6	−22.47
B	−23.96	−24.01
C	−20.11	−18.43
D	−23.84	−21.49
E	−20.92	−20.72

The level of the audio source contents should also be aligned so that the perceived loudness differences between the stimuli is small. Table 3.24 shows the RMS values of the stimuli in the TVM case. The RMS level values are the overall RMS values for the whole file. The files were processed with Adobe Audition 3.0.

For P.NAMS, a Matlab script is used for adjusting the RMS level of the audio source contents to approx. −26 dB. This RMS level is then "manually" adjusted for obtaining the same perceived loudness, based on informal tests with experts.

Note that the subjects had the possibility to correct the reproduction level individually before they started the actual test, if they felt the level was not adequate. In the case of the TVM tests, no subject wanted to change the listening level. In P.NAMS, only one subject decided to slightly lower the listening level.

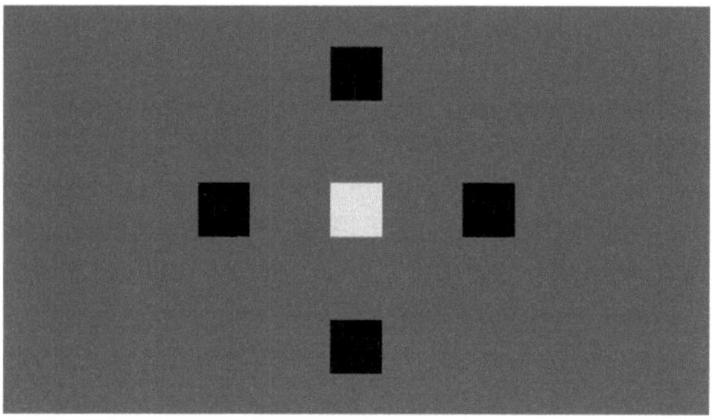

Fig. 3.8 Picture used for calibrating the screens

3.3.4.2 Viewing Environment and Video Calibration

The TVM tests were conducted at different locations, implying different test rooms and HD screens. For the TVM and P.NAMS tests conducted at T-Labs, up to two subjects participated in the test in parallel, in two different rooms. In each room, a high-quality 42-inches native HD (1920 × 1080) LCD display (Barco LC-42) was used. In all rooms, D65 (6500 K) lights were used as background light. The light was falling behind the screen, yielding a homogeneous background luminance of approx. 20 cd/m^2. The overall background illumination was below 20 lux. The gamma of the screens was equal to 2.2. The peak luminance of the screens was measured on the white square of Fig. 3.8, by forcing the R, G, B values of the screen to maximum values (i.e., 255). Typical measured peak luminance values were ranging between 180 cd/m^2 and 220 cd/m^2, falling in the range specified by ITU-R BT.710 (150–250 cd/m^2) for HD. This results in a ratio of luminance of background behind picture monitor to peak luminance of picture always smaller than 0.2, as specified in ITU-T P.910. Lastly, the color temperature of the screen was set to 6500 K. All light measurements were obtained with a Sekonic L-758 Cine light meter.

Note that the SD and HD720 tests were conducted on the native 1080p Barco screen as well. No up-scaling of the picture was applied in these cases. Instead, a mid-grey frame was displayed around the SD or HD720 picture.

3.3.4.3 Participants and Test Session

Between 23 and 29 subjects participated in each test, and each subject was allowed to participate in only one test (audio, video, or audiovisual).

Participants were instructed about the procedure of the test. They were then screened for visual acuity and color blindness, using the Wenzel plates and Ishihara test. If they did not fail the tests, the participants had to read the test instruction and do a training in which they saw/listened to all source contents, all degradation types, and the whole quality range they would see/listen to later in the actual test. At this stage, they would get used to the quality task and were allowed to ask questions to the supervisor, who was staying in the test room for the training phase only.

The participants were instructed by the assessment tool when to take a break, typically after 20–30 min. The whole test session was lasting no longer than one hour. Participants were not video or audio experts and were paid for their participation.

For all database sets, in order to limit the context effect, the presentation order of the test stimuli was randomized across subjects, and sub-sequent appearances of the same content were avoided. In the case of TVM, different randomization orders were used for each subject. In the case of P.NAMS, the subjective tests were specified to use a minimum of four randomized subject orderings.

3.3.4.4 Test Method

An Absolute Category Rating (ACR) was used for collecting subjective quality judgements. This method has been selected for its high efficiency (many files can be rated in one test session) and good repeatability (see Sect. 2.3.1). This method is less sensitive than test methods in which the degraded file is directly compared to a non-degraded reference, but it is also more realistic, since users do not have access to the reference signal at home. Indeed, subjects should not distinguish quality levels they would not distinguish at home.

For TVM, the uncompressed original audio and video were used as hidden references in the tests, but the scores for the hidden reference were not subtracted from the scores, that is no hidden-reference removal was applied. For P.NAMS, a high quality version of the audiovisual stimuli were used as reference instead of the uncompressed signals. This allowed P.NAMS participants to use the corresponding PCAP files for modeling, while keeping a high quality version of each source content. As for TVM, no hidden-reference removal was applied in the analysis of the P.NAMS tests.

3.3.4.5 Rating Scale

In the **TVM** tests, the subjects rated the quality using the continuous 11-point quality scale recommended in ITU-T Rec. P.910 (ITU-T Recommendation P.910 1999) and shown in Fig. 3.9a. With its half segments above "ausgezeichnet" and below "schlecht", this scale has the advantage of attenuating the saturation effect and of compensating for the fact that subjects avoid the extreme ends of the scale. The use of a continuous scale and numbers also make the scale closer to an interval scale (see Chap. 2, Sect. 2.3.2).

In the **P.NAMS** sets, the subjects rated the quality using the categorical 5-point quality scale, recommended in ITU-R Recommendation BT.500-12 (2009) and represented in Fig. 3.9b. The language used for the labels of the scale corresponds to the native language of the subjects. The choice of the five-point categorical quality scale results from a compromise between the P.NAMS/P.NBAMS participants. The main arguments in favor of this scale is that it is widely used in the telecommunications area, and thus potentially better understood by the telecom community. It should be noted that for the addressed application and degradation types, the 11-point and the

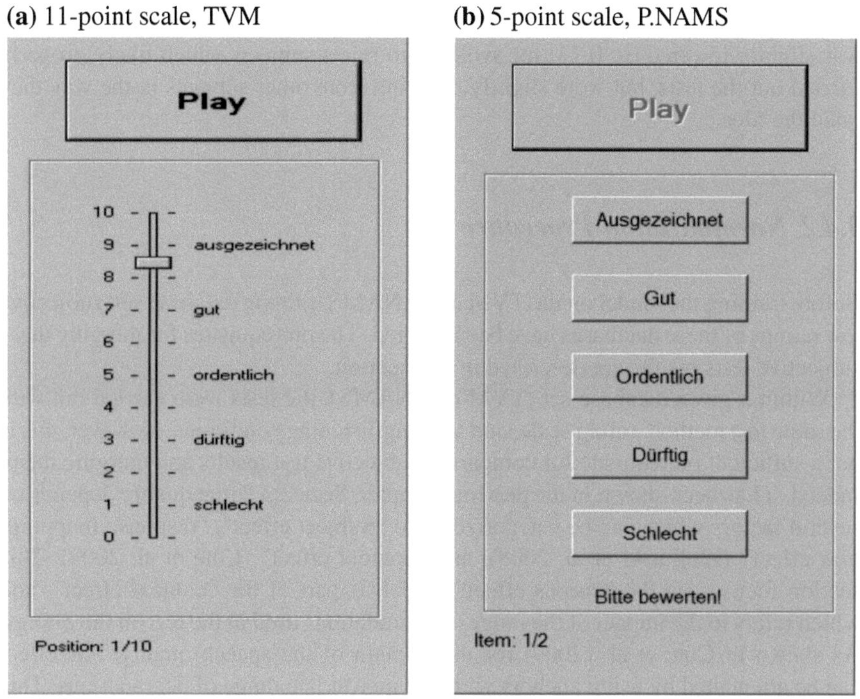

Fig. 3.9 Screenshot of **a** the 11-point and **b** 5-point quality scales (ITU-T Recommendation P.910 1999; ITU-R Recommendation BT.500-12 2009) used in the audio, video, and audiovisual subjective TVM and P.NAMS tests respectively. Attributes "schlecht", "dürftig", "ordentlich", "gut", "ausgezeichnet" correspond to "bad", "poor", "fair", "good", and "excellent" in the English version of the scale

5-point scales yield similar results in terms of repeatability and accuracy (Huynh-Thu et al. 2011).

3.4 Subjective Test Analysis Procedure

3.4.1 Rejection Criterion

For the TVM databases, the distribution of the ratings has been studied per subject, for detecting any unwanted rating behavior. For instance, it was expected that the subjects use the full range of the scale. As a consequence, if one subject used only a quarter or half of the scale, the ratings for that subject were rejected.

In the case of the P.NAMS databases, the ratings of one subject were rejected if the Pearson Correlation Coefficient between the ratings of this subject and the ratings averaged over all subjects was smaller than 0.75 for that database. With the analysis of the results, and especially of the distribution of the ratings, this threshold was slightly lowered (to 0.73) for avoiding to reject subjects which likely properly carried out the tests, but were slightly differing from other subjects in the way they rated the files.

3.4.2 Normalization Procedure

Before training the model on the TVM and P.NAMS training database set, subjective test ratings of these databases have been merged. The prerequisites for merging these subjective tests results are described in this section.

Within a given database set (TVM or P.NAMS), the tests were carried out with the same test method, rating scale, and viewing/listening conditions. However, this is not a sufficient prerequisite for comparing subjective test results and merging them. Indeed, it has been shown in the previous chapter, Sect. 2.3.2 that quality depends on several factors which can be categorized into "Subject effect", "response-mapping-bias effect" (Zieli´nski et al. 2008), and "context effect" (Côté et al. 2009). This section focuses on the "corpus-effect", which is part of the "context-effect", and which refers to the impact of the range of degradations used in the test on the ratings. As shown in Côté et al. (2009) for the domain of the speech quality, this effect can be attenuated by using anchors with the appropriate range of degradations. This motivated the use of anchors in all databases, with the anchors being discussed in section "TVM Test Design" and "P.NAMS Test Design".

In Côté et al. (2008); Möller et al. (2006), the authors show that the introduction of bad-quality conditions in a test leads to higher ratings of all other conditions in the test, and vice versa. This is referred to as the "positive" and "negative" corpus effect. It is also observed that transmission errors and background noise are dominant degradations, and they likely draw the attention away from speech codec distortion.

The positive corpus effect described in [CÔ8, Möl06] and summarized in Sect. 2.3.2 probably occurs in the tests conducted during this work since they contain more transmission error conditions than encoding-only conditions. This was intended because the tests are used for developing a model for the quality impact due to transmission errors. However, since the duration of the subjective tests should be kept short for not tiring the subjects, a limited number of conditions can be used per test, and the proportion of conditions with transmission errors is higher than in real IPTV services. Since transmission error conditions are dominant over compression conditions, this is assumed to lead to a "positive" corpus effect. In other words, compression conditions might be rated higher than in a corpus with less bad quality conditions.

The "Ref." test (Table 3.6) and the anchors presented in Table 3.7 are used for overcoming this effect. In particular, the "Ref." test has been designed for containing the widest range of typical IPTV degradations, with a proportion of conditions per degradation type representative of IPTV services, yielding a balanced perceptual environment for the IPTV situation. The anchor files have also been selected for spanning the complete quality range. Rating values per anchor files in the "Ref."

test are thus considered as the ground truth, i.e., the values the ratings of the anchor files in the other tests should target to. This procedure differs from the one followed by Pinson in Pinson and Wolf (2008). Indeed, instead of considering one test as the reference one, Pinson considers all tests as equal. For each anchor file, Pinson computes the "grand mean" over all tests and the mapping of each anchor file is done on this "grand mean". This procedure is justified if no test is considered more perceptually balanced and better covers the range of quality than the others, as is however the case for the "Ref" test.

In the case of the video TVM databases (see Tables 3.1 and 3.6), the subjective ratings are normalized as follows: subjective ratings are first averaged over all subjects per video file. Then, a linear mapping is computed for each test between the ratings of Tests 1–5 and the ratings of the "Reference" test, using only the anchor files. This linear mapping function is determined using least-square regression, and the first-order polynomial function shown in Eq. (3.1):

$$y_{ref} = a_N \cdot x_N + b_N \tag{3.1}$$

where the x_N values are taken from Test N with $N \in \{1, 2, 3, 4, 5\}$, the y_{ref} values are taken from the "Ref." test, and the regression coefficients a_N and b_N are test-specifics.

The resulting mappings are used for transforming the ratings of all files of Tests 1–5.

In the case of the two TVM audio tests, the ratings of the anchor files were matching without applying any polynomial transformation. No normalization was thus applied.

The first TVM audiovisual test conducted for this work contained some unwanted processing errors. As a consequence, the second TVM audiovisual test was taken as the reference test, and the same normalization procedure as for the TVM video tests was followed.

None of the P.NAMS training databases can be considered as a reference test. As a consequence, a normalization procedure close to the one of Pinson and Wolf (2008) was followed per modality (audio, video, and audiovideo). However, the "grand mean" was not computed from the anchor ratings but manually selected so that the min and max values of the anchor files (averaged over all subjects) were similar for all databases. Care was also taken for obtaining in each database the ratings that span the whole quality scale.

For each database, anchor files are mapped to the manually selected grand mean using Eq. (3.1). This time, the x_N values are taken from each P.NAMS tests, and the y_{ref} values are the manually selected grand mean, and the regression coefficients a_N and b_N are test-specifics. The respective mapping terms are used for transforming the ratings of all files in each P.NAMS test.

The P.NAMS validation set has been developed for evaluating the P.NAMS models. As a consequence, there is no need to merge the results of the validation set with the ratings of the TVM and P.NAMS training set. However, as explained in the following section, and for compensating the bias between the training and validation tests, a first-order polynomial mapping will be applied on the output of the models and taking as target values the subjective ratings averaged per stimulus over all subjects.

3.5 Model Development

3.5.1 Modeling Approach

For modeling audio, video, and audiovisual quality, the impairment-factor based approach presented in Sect. 2.2.3 and based on the findings by Allnatt (1983) will be followed. The models are thus expressed with the following Eq. (3.2):

$$Q_X = Qo_X - \sum_{k=1}^{N} IF_X^k \tag{3.2}$$

where Q_X is the audio ($X \equiv A$), video ($X \equiv V$), or audiovisual ($X \equiv AV$) quality. Qo_X is the base audio, video, or audiovisual quality, and takes as value the maximum value of the modeling scale. N is the total number of impairment factors IF, and IF_X^k is an Impairment Factor for modality X with indices k.

Two impairment factors are used for the audio and video quality model, one related to the quality impact due to compression (coding) artifacts: $IF_X^1 = Icod_X$, and one related to the quality impact due to transmission errors: $IF_X^2 = Itra_X$, yielding Eq. (3.3). $Icod_X$ and $Itra_X$ correspond to the two perceptual quality dimensions discussed at the end of Sect. 3.2 and in Fig. 3.2.

$$Q_X = Qo_X - Icod_X - Itra_X, \tag{3.3}$$

The impairment-factor based audio and video quality models are described and evaluated in Chaps. 4 and 5.

For the audiovisual quality model, the respective integration corresponds to the audio and video impairment factors of the audio and video quality models, but also to the combination of these impairment factors (Garcia et al. 2011), yielding Eq. (3.4).

$$\begin{aligned} Q_{AV} = \ & Qo_{AV} \\ & - b \cdot Icod_A - c \cdot Icod_V \\ & - d \cdot Icod_A \cdot Icod_V \\ & - e \cdot Itra_A - f \cdot Itra_V \\ & - g \cdot Itra_A \cdot Itra_V \\ & - h \cdot Icod_V \cdot Itra_A - i \cdot Icod_A \cdot Itra_V \end{aligned} \tag{3.4}$$

This IF-based audiovisual quality model will be detailed in Chap. 6. It will also be compared to the more traditional quality-based model for which the inputs are the audio (Qa) and video (Qv) qualities, as shown in Eq. (3.5).

$$Qav = \alpha + \beta \cdot Qa + \gamma \cdot Qv + \zeta \cdot Qa \cdot Qv. \tag{3.5}$$

3.5.2 Model Training

All models are developed based on the results of the subjective tests described in Sect. 3.3. For each database, they are averaged over all subjects, i.e., per stimulus. For the TVM tests, which use the 11-point P.910 scale for rating, the scores were averaged over all subjects, i.e., per stimulus, yielding Mean Opinion Scores (MOS_{11}), then linearly transformed to the 5-point ACR MOS scale (MOS_5) by aligning the numbers of the scales (see Eq. 3.6), and further transformed to the 100-point model-scale (Q_{100}) using the conversion defined in ITU-T Recommendation G.107 (2005), and shown in Appendix I and Fig. 3.10.

$$MOS_5 = MOS_{11} \cdot \frac{4}{10} + 1; \tag{3.6}$$

Following Allnatt's approach (1983), the 100-point scale, also referred to as "R-scale" in ITU-T G.107, is assumed to be a psychological scale, i.e., a scale on which impairment factors are expected to be additive. Indeed, with this transformation, which follows an S-shape, a small difference at the ends of the MOS scale is expended on the 100-point scale. This transformation thus compensates for the "saturation" effect occurring with the use of the 5-point MOS scale (see Sect. 2.3.2). Note that this effect should be smaller with the 11-point scale, thanks to the additional half-interval at the ends of the scale. This is not proven by respective analyses given, e.g., in Huynh-Thu et al. (2011). It is unclear if the ITU-T G.107 transformation should be used on the TVM data, for which ratings have been collected on the 11-point scale. However, modeling has been tested both using the MOS scores and the scores transformed on the 100-point scale (Q_{100} scores). Model performance results using the Q_{100} scores were at least as good as when using the MOS scores. It has therefore been decided to keep the ITU-T G.107 transformation for transforming the TVM scores to the 100-point model-scale.

For P.NAMS, which uses the 5-point ACR MOS scale, the averaged ratings were directly transformed onto the 100-point scale using the ITU-T G.107 transform.

Now that the model form has been selected (Eq. 3.2), the next modeling task consists in choosing the most appropriate variables (input parameters) and mapping functions for estimating the selected impairment factors $Icod$ and $Itra$.

Multiple regression analysis was used as common modeling method for all models. More details on the modeling will be given in the chapters dedicated to each model. Similarly, more details will be given on the selection of appropriate

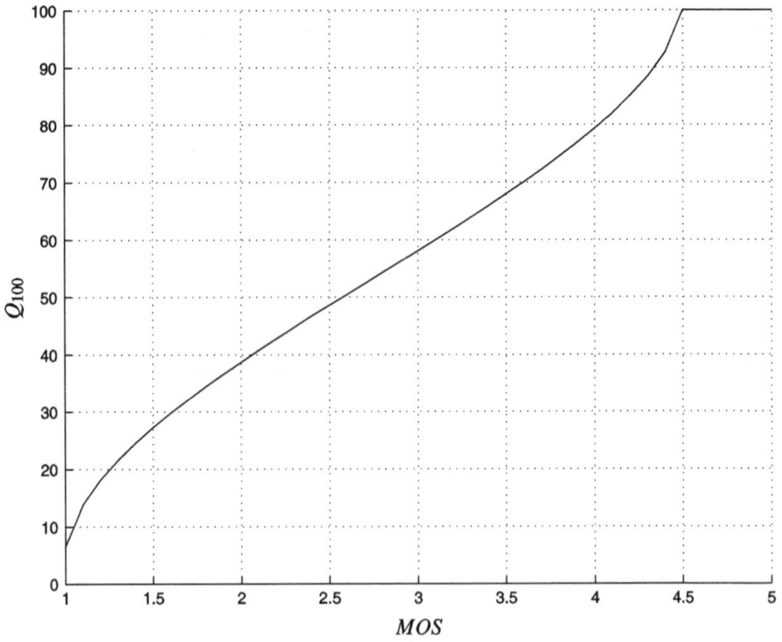

Fig. 3.10 Transformation of MOS scores onto the psychological 100-point model-scale (Q_{100}) using the equations defined in ITU-T G.107 and reproduced in Appendix I

parameters, the choice of parameters being restricted due to the encryption of the audiovisual stream.

It should be noted that not all database sets were available at the same period of time. Indeed, the TVM set was first developed and extensively used for analysing the quality impact due to the different degradations. A first version of the audio, video, and audiovisual quality models was developed from the TVM database set. Once the P.NAMS training set became available, the models were further developed and re-trained using the P.NAMS training sets and partially the TVM sets. More details will be given for each model (audio, video, and audiovisual) in the dedicated chapters.

3.5.3 Model Evaluation

At this stage, i.e., before the P.NAMS validation database sets was available, a cross-validation procedure is applied for each model (audio, video, and audiovisual) in order to evaluate the model. The RMSE (see Sect. 2.4.3) is used as evaluation criterion, as defined in the P.NAMS requirements. Since it is often used in the literature, and because it provides a measure of the monotonicity of the model, the Pearson

Correlation Coefficients is also provided. The *RMSE** is computed as well for comparison purpose with other studies. Indeed, by including the confidence interval of the subjective ratings in its computation, the *RMSE** of models evaluated on different databases—which most probably have different ranges of confidence intervals—become comparable, provided that the same scale has been used for creating the different databases. More details on the cross-validation procedure will be provided in the "Model Evaluation" section of Chaps. 4, 5, and 6.

Each model (audio, video, and audiovisual) is ultimately evaluated against unknown data, the P.NAMS validation database set, once available. The RMSE is again used as evaluation criterion, and the Pearson Correlation coefficients and *RMSE** are provided as additional information. As described in Sect. 2.4.3, each metric (RMSE, Pearson Correlation coefficients, *RMSE**) is computed per experiment. To obtain the overall performance of the model, the output of each metric is averaged over all validation databases. Since these databases correspond to the same modality (audio, video, or audiovisual) and are used for the same validation purpose (and not for training the model), equal weights are attributed to all validation databases.

The scores outputted by the audio, video, and audiovisual quality models presented in this work are expressed on the 100-point model-scale. Since the performance indicators mentioned above compare the quality scores outputted by the models with the ones from the subjective tests, the output of the models is first transformed back to the 5-point-MOS scale used in the subjective tests. This step is done by applying the "MOS from Q_{100}" transformation provided in Appendix I, and corresponding to the inverse transform of the ITU-T G.107 transform. For compensating for potential bias between each subjective test, a first-order polynomial mapping is applied in addition on the transformed output of the models, mapping to the subjective ratings that are used as target values (averaged over all subjects). The mapping was applied per database for each of the P.NAMS validation database sets.

As will be detailed in Chaps. 5 and 6, several video and audiovisual quality models have been developed in this book. For comparing their performance, the significance of the difference of the RMSE and of the Pearson correlation coefficients of the different models is computed (see Sect. 2.4.3).

In order to better assess the performance of a given model, other models from the literature are generally compared to the given model. In this work, some of the parametric models proposed in the literature, like the audio and video quality models from NTT (Egi et al. 2010; Yamagishi and Hayashi et al. 2008), were implemented during the development of the model proposed in this book (the "T-V-Model"). They were used as baselines against which the T-V-Model was compared to. However, it is assumed that, for thc purpose of the P.NAMS competition, NTT, and of course the other competitors, have further improved their models. These models, due to the competition, were not published at the time of completion of this book, and therefore could not be implemented. However, their performances, including the T-V-Model performance, were compared officially by the ITU-T Q.14 group during the P.NAMS competitions. Since the T-V-Model won the P.NAMS competition for the higher resolution application area (IPTV, VoD), and is now standardized, it is guaranteed that the model achieves high performance, especially in comparison with

Table 3.25 Comparison of the TVM, P.NAMS training and validation database sets

	TVM	P.NAMS
Num audio DBs	2	Training: 2; Validation: 2
Num video DBs	12	Training: 5; Validation: 8
Num AV DBs	4	Training: 3; Validation: 4
Resolution	SD (PAL), HD1080	SD (PAL, NTSC), HD720, HD1080
Audio format	48 kHz, 16 bit, stereo	44.1 kHz, 48 kHz, 16 bit, mono and stereo
Frame-rate	24 fps, 25 fps	25 fps, 30 fps, 50 fps, 60 fps
Scanning mode/DB	Interlaced and progressive	Interlaced or progressive
Content	5 contents/DB, 16 s each	8 contents/DB, 10 s each
Video codec	MPEG-2 main profile and H.264 main profile	H.264 main and high profiles
Audio codec	Dolby AAC-LC, Dolby HE-AACv2, MP2, MP3	Nero AAC-LC, Nero HE-AACv2, MP2, AC3
CBR vs. VBR	CBR	CBR
Encapsulation	No PCAP	PCAP-based
AV mux	None	TS-Multiplexed
Packet loss	Random and bursty (2 state-MM)	Random and bursty (4 state-MM)
Slicing configuration	1 slice/MB-row and 3 slices/frame	Training: 1 slice/MB-row or 1 slice/frame; Validation: 1 slice/MB-row and/or 1 slice/frame
Viewing/listening environment	LCD, loudspeakers (stereo)	LCD, headphones (stereo)
AV asynchrony	None	None
Test method	ACR	ACR
Rating scale	11-point ITU P.910 scale	5-point quality scale
Modeling scale	100-point scale	100-point scale
Usage	IPTV degradation analysis, 1st model version	Training DBs: training + model enhancement; validation DBs: validation

DB DataBase
Num Number of
CBR Constant BitRate
VBR Variable BitRate
AV Audiovisual
MM Markov-Model
MB MacroBlock

the state-of-the-art models proposed by the other participants (more information on the model performance results can be found in ITU-T Recommendation P.1201 (2012)). As a consequence, models from the literature will generally not be compared in this work in terms of model performance to the T-V-Model. However, the model equations and input parameters proposed in the literature will be carefully studied and will serve as baselines for improving the T-V-Model.

3.6 Summary

All required choices for running subjective tests and, based on the subjective test results, developing the audio, video, and audiovisual quality models for IPTV services have now been made.

Three database sets have been produced: the TVM set and the P.NAMS training and validation sets. All databases have the same scope and are targeting the same application. They however differ in several points, the main divergences being between the TVM and P.NAMS sets. The P.NAMS sets were indeed produced months later, which provided the opportunity of using more advanced processing tools. They also result from compromises between the P.NAMS competitors. Table 3.25 summarizes the common points and differences between the three database sets.

The TVM set was mainly used for better understanding the quality impact due to IPTV degradations and producing a first version of the models. The P.NAMS training set was produced for improving the models and adapting them to the P.NAMS competition requirements. A cross-validation procedure was undertaken on the TVM and P.NAMS training sets for preventing overtraining. The models were eventually evaluated on the P.NAMS validation set.

The following three chapters analyse the subjective test results and present and validate quality models for the respective audio, video, and audiovisual components.

A single audio quality model will be developed for both the network planning and quality monitoring cases (Chap. 4), while one video quality model will be developed for each case, planning and monitoring (Chap. 5). At least, two modeling approaches will be tested for the audiovisual quality integration model and for both the planning and monitoring application, resulting in four audiovisual quality models, which will be validated and compared in Chap. 6.

Chapter 4
Audio Quality Model

This chapter presents the parametric audio quality model developed in this work. This audio quality model is trained on subjective ratings collected from the audio quality tests presented in Sect. 3.3. The chapter starts with a short reminder on the selected test methods and test design before detailing specific subjective test results. An overview of existing parametric audio- or speech-quality models is then provided before presenting the model developed based on the test results. Due to the systematic, modeling-directed approach used in the TVM database set, the TVM results serve as basis for the data analysis. The P.NAMS training data set (see Sect. 3.3.3) is used to confirm the conclusions drawn from the TVM data set in terms of modeling, and to extend the analysis to other codec types and implementations. The audio quality model presented in this chapter is similar to the model submitted to the P.NAMS competition. The ultimately submitted model contains a second, back-up module to handle specific cases of conditions. The model variant described in this work is compared to the model submitted to the competition. Ultimately, it became clear that both model variants are performing equally well.

The model presented in this work has been optimized on the P.NAMS training database set and evaluated following the P.NAMS validation procedure. As a first step, and because the P.NAMS validation set was not available during the model development, a cross-validation procedure is applied on the P.NAMS training set to check that the model is not overtrained. The model is then validated against unknown data using the P.NAMS validation database set once available. The TVM database set is used for evaluating the model performance on codec type and implementation not covered by the P.NAMS scope.

The model shows high performance results on unknown data with a Pearson correlation coefficient of 0.92 with the subjective quality scores and an RMSE of 0.39 on the 5-point scale. These performance results are equal to those of the audio module of the ITU-T P.1201.2 standardized model, which is expected since the audio model presented in this chapter served as the basis for the audio module of the P.1201.2 standard.

M.-N. Garcia, *Parametric Packet-Based Audiovisual Quality Model for IPTV Services*, T-Labs Series in Telecommunication Services, DOI: 10.1007/978-3-319-04855-0_4,

4.1 Subjective Quality Assessment

4.1.1 Subjective Test Overview

The audio quality subjective tests have been aligned to the video quality subjective test methods. For collecting opinion ratings, the audio quality tests were conducted using an ACR test method using the continuous 11-point scale (Huynh-Thu and Ghanbari 2005) in the TVM database sets, and the 5-point categorical scale in the P.NAMS training and validation database sets, as described in detail in Sect. 3.3.4. Two audio tests were conducted in each database set.

As described in Sects. 3.3.2 and 3.3.3, the same anchors (Tables 3.9 and 3.22) were used in all subjective tests for a given database set. These anchors cover the whole quality range and degradation types used in the tests. They are used for mapping the results of both tests, as described in Sect. 3.4.2.

The audio contents used in the tests are listed in Sects. 3.3.2 and 3.3.3, Tables 3.4, 3.13, and 3.14.

In order to insert typical IPTV degradations, each audio source sequence was processed off-line following the test conditions described in Table 3.8 for the two TVM databases, and in Tables 3.15 and 3.18 for the four P.NAMS databases.

More information on the test design and processing chain was given in the previous chapter, Sects. 3.3.2 and 3.3.3.

For the TVM databases ($ph2_A$ and $ph3_A$), this results in 51 ($ph2_A$) and 40 ($ph3_A$) audio test conditions for each of the five audio contents, leading to 255 ($ph2_A$) and 200 ($ph3_A$) audio sequences to be rated by the subjects. For each of the P.NAMS databases apart from $tr16$, 30 test conditions were processed with eight contents, resulting in 240 audio files per P.NAMS database. For $tr16$, since one content was removed from the set of eight contents (see Sect. 3.3.3, "P.NAMS Audiovisual content"), 34 test conditions were processed with seven contents, resulting in 238 audio files.

4.1.2 Subjective Test Results

Results were collected from the TVM audio databases ($ph2_A$ and $ph3_A$) and from the two P.NAMS audio training databases ($tr07$ and $tr16$). Ratings of the P.NAMS database $tr07$ are not displayed, for IPRs reasons[1], but they were used for model training and further development.

Based on the discussion of Sect. 3.4.2, the ratings of the anchor files were matching between $ph2_A$ and $ph3_A$, and the results of the two databases can be merged without applying any polynomial transformation on the ratings. Similarly, no mapping was needed between the $tr07$ and $tr16$ ratings. As will be explained in section "Quality

[1] The entire database set for P.NAMS was produced under NDA between the participants of the P.NAMS and P.NBAMS competitions.

impact due to audio compression", and based on the analysis of the subjective test results for the only codec used in both data sets with the same implementation (MP2), no mapping was required either between the TVM and P.NAMS training audio tests.

For the TVM subjective tests, the quality ratings were averaged over all subjects and error patterns in the case of audio frame loss, as stated in Sect. 3.3.2.2. The obtained averages were first linearly transformed to the five-point MOS scale, then to the 100-point model-scale using the transformation defined in ITU-T G.107 (ITU-T Recommendation G.107 2005), Appendix I, and reproduced in the Annex I of this book. For the P.NAMS subjective tests, quality scores were averaged over all subjects, then transformed from the five-point scale to the 100-point scale, using the same transformation. In the case of the P.NAMS data set, only one error pattern was used per combination of content and conditions with frame loss.

Note that when results from a single database set are displayed, the original scale used in the tests (11-point scale for TVM, 5-point scale for P.NAMS) are used for the plots. This enables the use of confidence intervals as obtained in the tests. When results from TVM and P.NAMS are merged, the 100-point model-scale is used in the graphs.

As mentioned in the introduction of this chapter, the TVM database is primarily used for analysing the results and for deriving respective modeling approaches. The test design and processing chain of the TVM database set allows a systematic analysis of the results, especially in the case of frame loss. Both random and bursty losses have intensively been tested, and the achieved frame loss rate and burst lengths have carefully been controlled. In addition, the use of four error patterns per combination of content and condition allows compensation from the influence of the loss location on perceived quality. As a complement, the results per error pattern averaged over subjects allow the influence of the loss location on the perceived quality to be analysed. The processing chain used for the P.NAMS database is more realistic than the one used for the TVM data set, but for the P.NAMS data, the actual frame loss rate and burst lengths vary from one content to another one, and only one error pattern was used per condition.

The analysis of the subjective test results is organized as follows: the first part focuses on the quality impact due to the different audio codecs when no audio frame loss occurs. The second part is dedicated to the quality impact due to audio frame loss, for both random and bursty loss, and for different audio codecs. Some of the TVM test results have already been presented in (Feiten et al. 2009).

4.1.2.1 Quality Impact Due to Audio Compression

Figure 4.1 shows the quality impact due to the audio bitrate for the five codecs used in the tests under error-free conditions (i.e., without packet loss). Results are averaged over all contents, to compensate for the influence of the content on quality. As expected from the respective audio standardisation work (ISO/IEC 11172–3 1993, ISO/IEC 13818–3 1995, ISO/IEC 13818–7 2006, ISO/IEC 14496–3 2006), the best

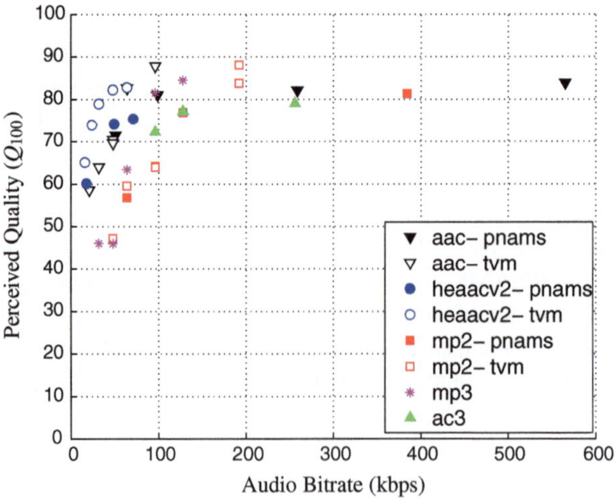

Fig. 4.1 Perceived audio quality, on the 100-point scale Q_{100}, for five different codecs and various bitrates. TVM and P.NAMS (*tr*16) database sets

performing codec is HE-AAC, followed by, in that order, AAC-LC, MP3, AC-3, and MP2. For each codec, the perceived quality shows an exponential dependence of the bitrate.

It should be noted that the maximum quality achieved in the P.NAMS test is lower than for the TVM test. This may reflect a stronger saturation effect in P.NAMS. Since results shown in Fig. 4.1 are averaged over all contents, some P.NAMS contents may be associated with a low base quality and therefore pull the average down. This assumption will however later be rejected when analysing the quality impact due to the content. This difference of behavior at high bitrates may be compensated by mapping the results between the two database sets. However, it can be observed in Fig. 4.1 that results for MP2 are mapping well between the two database sets. This observation will be confirmed later in the case of frame loss. Since MP2 is the only codec used with identical implementation between the two database sets, and hence differences between tests may be codec implementation-dependent, it has been decided not to use any mapping between the two database sets.

As shown in Fig. 4.1, the quality ratings for both HE-AAC and AAC-LC are higher in TVM ("heaacv2-tvm" and "aac-tvm") than in P.NAMS ("heaacv2-pnams" and "aac-pnams"). This results may either be due to the higher saturation effect observed in P.NAMS (see above), or to a better implementation of the HE-AAC and AAC-LC codecs in TVM (Dolby implementation), than in P.NAMS (Nero implementation). Both implementations yield higher quality than the other codecs at low and medium bitrates.

For the no-loss case and the TVM databases, an analysis of variance using the perceived quality as dependent variable, the codec, bitrate, and content as independent fixed factors, and the subjects as random factor, was computed, and a significant effect

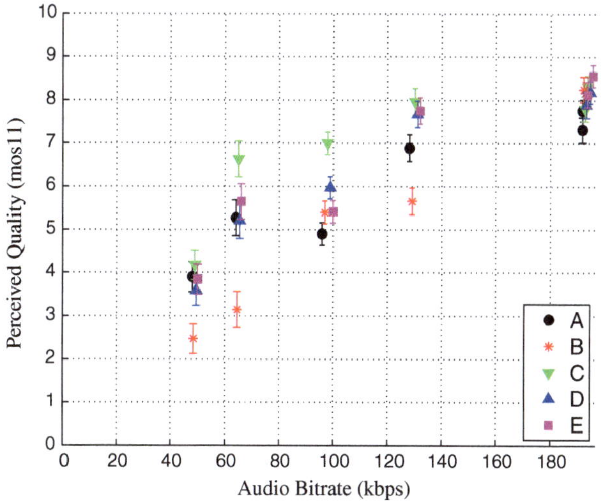

Fig. 4.2 Perceived audio quality (on the 11-point (*mos*11) scale) as a function of the audio bitrate for various content types (see Table 3.4) and for the MP2 codec. TVM database set

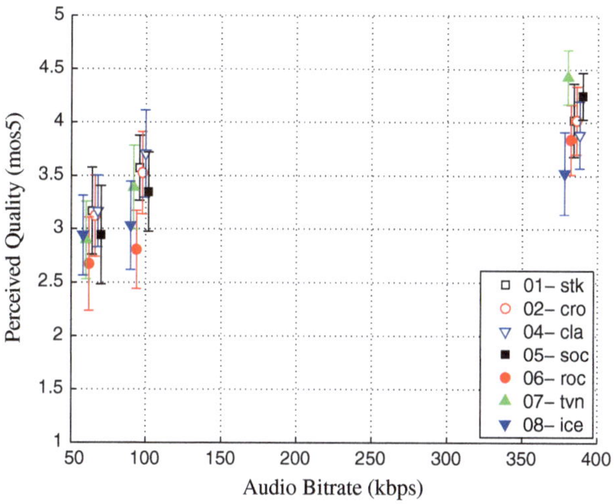

Fig. 4.3 Perceived audio quality (on the 5-point (*mos*5) scale) as a function of the audio bitrate for various content types and for the MP2 codec. P.NAMS (*tr*16) database set

of the content on the perceived quality was found. This is particularly valid for MP2, as shown in Fig. 4.2, for which the speech-only content (content "B", see Table 3.4) yields lower quality than the other contents. Results are displayed on the 11-point scale (*mos*11) used in the subjective tests so that the confidence interval of the ratings averaged over all subjects can be shown as obtained in the tests.

This latter behavior does not seem to apply to the P.NAMS databases. This is especially true for the $tr16$ database, where the speech-only content does not present lower quality than the other content types at low and medium bitrates, as shown in Fig. 4.3. Similarly, no content dependency could be observed for any of the other codecs used in the P.NAMS database sets. However, it should be noted that all of the P.NAMS audio contents are of a more complex nature than the more archetypical TVM contents. They mix different content types (speech, singing, crowd noise, ringing bells, gun shot), leading to more similar contents, and reducing the content impact on quality. In the case of encrypted data and packet-header based models, audio content-related parameters can anyways not be extracted from the bitstream, and therefore not be used in the model. If no information on the content type is sent to the model as side information, the model is constrained to predict the same quality score for all content types.

4.1.2.2 Quality Impact Due to Audio Frame Loss: Random Loss

Figure 4.4 shows the quality impact due to random audio frame loss for four codecs (HE-AAC (Dolby implementation), AAC-LC (Dolby implementation), MP2 and MP3) at various bitrates, and based on the TVM scores.

The P.NAMS databases focus on bursty IP-packet losses, resulting in non-random audio frame loss, and are therefore not used for the analysis of the quality impact due to such random audio frame loss. It should be noted that for the TVM data, the term "random" indicates that one and only one frame has been lost in a row. In theory, more than one frame can be lost in a row in the case of random loss.

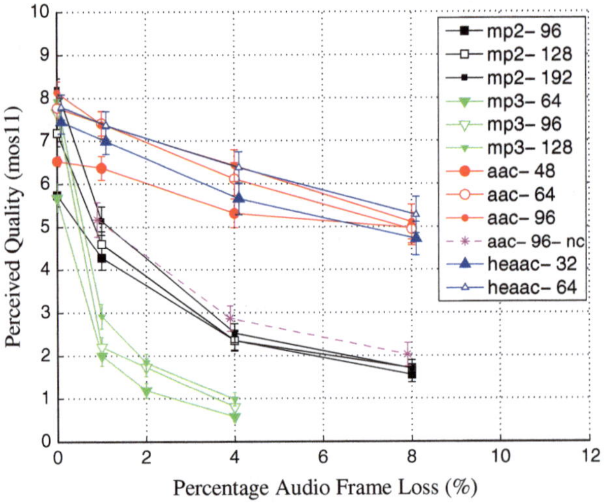

Fig. 4.4 Perceived audio quality for random audio frame loss for different codecs and bitrates (32–192 kbps); *nc*: concealment is switched-off; TVM database set

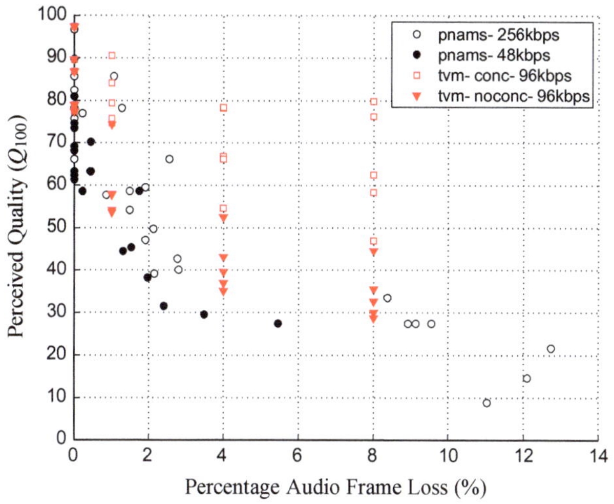

Fig. 4.5 Perceived audio quality for AAC-LC in the case of audio frame losses, with (*conc*) and without (*noconc*, *pnams*) concealment. TVM and P.NAMS (*tr*16) database sets

It can be observed that for a given bitrate, HE-AAC (v2) and AAC-LC yield similar quality under packet-loss. HE-AAC seems to be performing slightly better than AAC-LC at equal bitrate (64 kbps), but the difference is not significant, although systematic. These results are expected since HE-AAC combines the AAC technology with additional tools [spectral band replication and parametric stereo (HE-AACv2 only)] for extending AAC towards bitrates lower than 128 kbps [see Sect. 2.2.1.3 and (Meltzer and Moser 2006)]. The reduced advantage of HE-AAC over AAC may be due to the loudspeaker-based listening applied in case of the TVM data set, and a related reduction of audio coding artifacts audibility (Schobben and van de Par 2004).

Both HE-AAC and AAC-LC codecs are much less sensitive to frame loss than MP2 and MP3. In turn, MP2 performs better than MP3. These results are expected since the employed HE-AAC and AAC-LC decoders apply interpolation-based and frame-repetition-based concealment, while the MP2 and MP3 decoders insert silence frames as error concealment. This observation is confirmed by analysing the quality impact due to audio frame loss when the Dolby AAC-LC loss concealment is switched off ("aac-96-nc" in Fig. 4.4), and silence is inserted in the case of frame loss. In that case, the AAC-LC quality drops to the MP2-level quality. Similar results are observed in the P.NAMS database set, where a Nero implementation of the AAC-LC codec has been used, which apparently uses silence insertion as loss concealment.

As can be observed in Fig. 4.5, at high bitrate (96/256 kbps), the Nero AAC-LC implementation ("pnams-256 kbps") yields similar quality as the Dolby AAC-LC implementation without concealment ("tvm-noconc-96 kbps"). Note that the highest bitrate used in TVM is lower than the highest bitrate used in P.NAMS. However, since the maximum quality is already reached at 96 kbps for the no loss case, see Fig. 4.1,

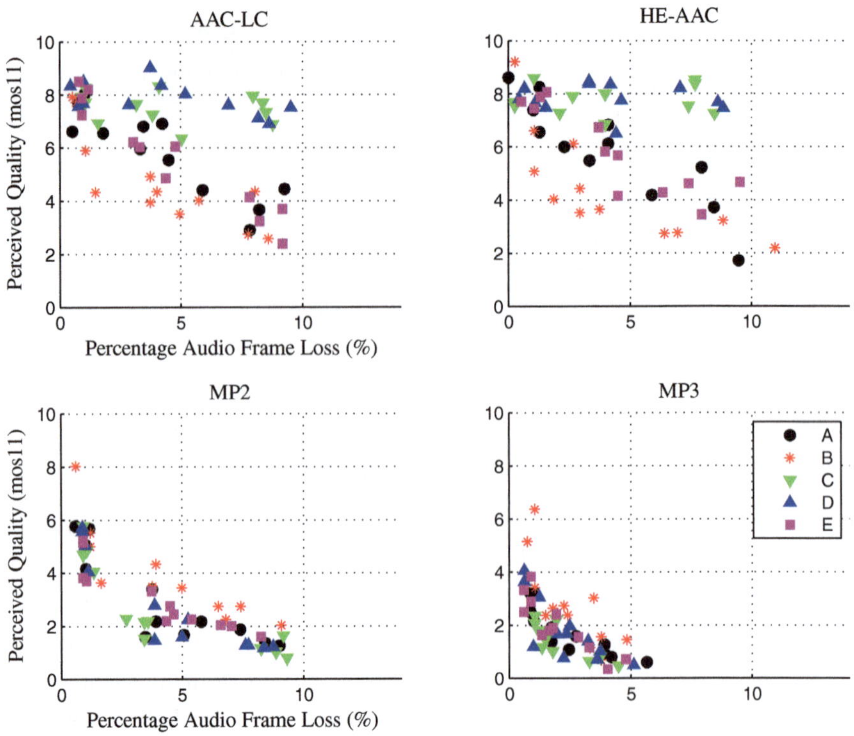

Fig. 4.6 Perceived audio quality for random audio frame loss for each codec and for high bitrates. Results are averaged per content and per error pattern. TVM database set

there are, in both cases, no compression artifacts which may have influenced the results.

Finally, for low to medium percentage of audio frame loss and all codecs, the perceived quality increases with the bitrate. This reflects the impact of compression in the case of low to medium percentage of audio frame loss while this effect is hidden or masked in the case of high percentage of audio frame loss.

In the case of audio frame loss, results were so far averaged over all subjects, contents, and error patterns. It is worth investigating the validity of the above observations when taking the results per content and per error pattern only averaged over all subjects.

Figure 4.6 shows the perceived quality in case of random loss for each codec, taking the results averaged per content and error pattern. It is obvious that the loss concealment algorithms implemented in HE-AAC and AAC-LC perform much better on contents not containing speech (contents "C" and "D"). This is expected since speech signals are much less stationary than non-speech signals. Therefore, when repeating an audio frame for compensating a lost frame as is the case with AAC-LC, or interpolating the audio frame from surrounding non-lost frames, as is the case with

HE-AAC, the concealment is much less audible in the case of stationary signals. This content dependency does not hold true for MP2 and MP3, where silence insertion was used as PLC. In contrast, the speech-only signal "B" seems to yield better quality than the other types of signals.

Two instantiations of the same percentage of audio frame loss for a given codec and bitrate are represented by the same symbol on a vertical line in Fig. 4.6. It can be observed from this figure that the influence of the error pattern and thus of the loss location is low, especially compared to the quality impact due to loss rate and content in the case of HE-AAC and AAC-LC.

Recall that only one fourth of the subjects rated each error pattern. As a consequence, the 95 % confidence interval values are higher than the ones usually obtained when taking the average over all subjects. This reduces the probability of having a significant difference between the perceived quality of two instantiations of the same percentage of audio frame loss, with the given number of subjects. For the sake of clarity, the 95 % confidence interval was not represented in Fig. 4.6.

An analysis of variance taking, as dependent variable the perceived quality, as fixed factors the condition, the content and the error-pattern, and as random factor the subject, confirmed that the influence of the content type on the perceived quality is significant but not the influence of the error-pattern.

4.1.2.3 Quality Impact Due to Audio Frame Loss: Bursty Loss

At this stage, the quality impact due to audio frame loss has been studied only in the case of random loss.

The quality impact due to bursty frame loss for MP2, the only codec tested under bursty losses in the TVM database set, is shown in Fig. 4.7. As for random loss, a small impact due to the bitrate is observed for the different burstiness levels, with higher quality for higher bitrate. The average burst length, represented in the figure by μ, ranges from one to six frames lost in a row. At high percentage of audio frame loss, the average burst length has no impact. At lower percentage of audio frame loss ($Ppl = 1\,\%$), two tendencies can be observed. First, random loss ("rand-") tends to yield higher quality than short burst-length losses ("μ_2-", i.e., $\mu = 2$). This probably reflects that isolated losses, due to their short duration, but also depending on their location, have a higher probably of not being perceived. Then longer burst length ("μ_4-", i.e., $\mu = 4$, and "μ_6-", i.e., $\mu = 6$) enhances the quality. This observation indicates that for a given bitrate and percentage of audio frame loss, longer burst lengths and thus lower amount of loss events are less annoying than short but perceivable more frequent losses. Note that selecting six frames lost in a row as maximum burst length was motivated by the frame encapsulation: the common case of having six audio frames in a PES (Packetized Elementary Stream) frame was simulated. When losing an audio frame in the PES frame, the decoder cannot re-synchronize till the next PES header. As a consequence, losing the first audio frame of the PES frame results in losing six frames in a row. In the case of audiovisual

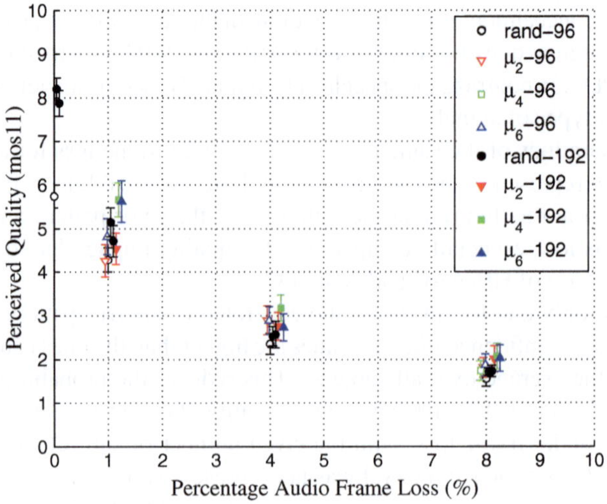

Fig. 4.7 Perceived audio quality for random and bursty audio frame loss for the MP2 codec and for two bitrates (96 and 192 kbps). *rand*: random frame loss; μ: average burst length; μ_x: $\mu = x$ (see text for details). TVM database set

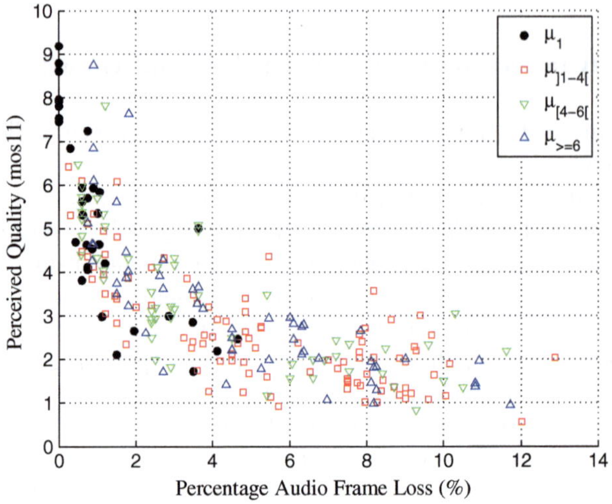

Fig. 4.8 Perceived audio quality for bursty audio frame loss for the MP2 codec and for high bitrates (192 kbps). Results are taken per error-pattern. μ: average burst length (see text for details). TVM database set

bitstreams, due to the much higher video bitrate compared to the audio bitrate, the probability of losing more than six frames in a row is almost zero and thus not tested.

Figure 4.8 shows the perceived quality as a function of the percentage of audio frame loss for high bitrate-encoded MP2 audio and bursty losses when taking the

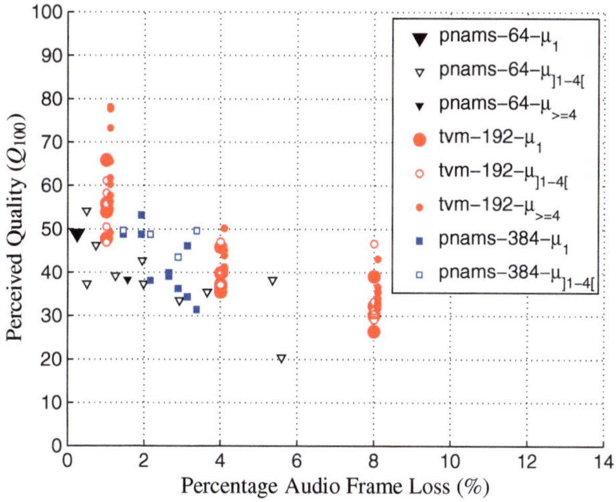

Fig. 4.9 Perceived audio quality for MP2 in the case of random and bursty losses, and for different bitrates (64, 192, 384 kbps). μ: average burst length (see text for details). P.NAMS ($tr16$) and TVM database sets

results per error-pattern. As before, μ represents the average burst length, i.e., the averaged number of consecutively lost frames. Four average burst length categories were used for classifying the results: audio files containing isolated single lost audio frames (μ_1); audio files with an average burst length between one (excluded) and four (excluded) ($\mu_{]1-4[}$); audio files with an average burst length between four (included) and six (excluded) ($\mu_{[4-6[}$), and audio files with an average burst length equal or longer than six ($\mu_{>=6}$). It has been observed that the burst length influences the perceived quality in the case of low percentage of audio frame loss. By now taking the ratings per error-pattern, it can be verified whether averaging the ratings per error-pattern is hiding a higher influence of the burst length. Figure 4.8 confirms the observation previously made on Fig. 4.7: the influence of the burst length, although quite low, is visible, especially for low percentages of audio frame loss.

For verifying whether the observation made on MP2 for TVM in the case of bursty loss is valid for P.NAMS, Fig. 4.9 shows the perceived quality as a function of the percentage of audio frame loss for different bitrates and bursty losses in the case of the MP2 codec for both the P.NAMS and the TVM database sets. For TVM, the results are averaged over the four error patterns, but are displayed per content. For P.NAMS, only one error pattern was used per combination of content and condition, and not the same percentage of audio frame loss was achieved for all contents for a given condition, which makes the graph less well aligned to the audio frame loss axis than for the TVM data. The average burst length also varies between contents for a given condition. As a consequence, three average burst length categories were used for classifying the audio P.NAMS files: files with isolated single lost audio frames (μ_1), files with an average burst length between one (excluded) and four (excluded)

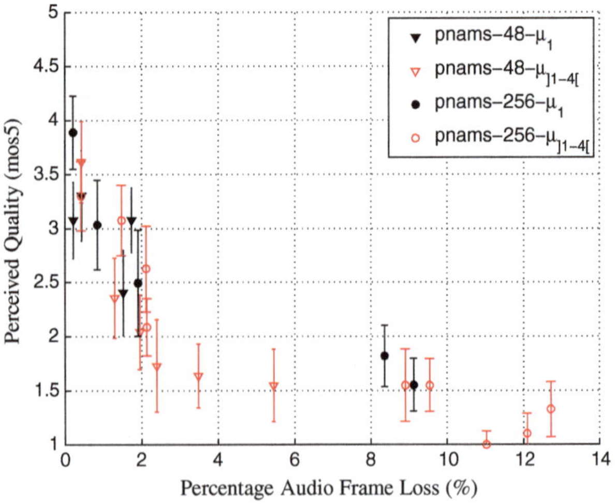

Fig. 4.10 Perceived audio quality for AAC-LC in the case of random and bursty losses, and for different bitrates (48, 256 kbps). μ: average burst length (see text for details). P.NAMS ($tr16$) database set

($\mu_{]1-4[}$), and files with an average burst length equal or longer than four ($\mu_{>=4}$). For the TVM data, which are this time displayed per content, Fig. 4.9 confirms that for bursty losses, an average burst length longer than four (third category) generally achieves higher perceived quality than audio files with identical percentage of audio frame loss and bitrates, but lower burst lengths. For the P.NAMS data, the burst length does not seem to impact quality. However, except for one case, all bursty cases have a burst length shorter than four, the threshold above which the impact of burst length becomes visible. As for the TVM dataset, the perceived quality due to audio frame loss tends to be higher at high bitrates ("pnams-384-") than at medium bitrates ("pnams-64-").

It must be noted that this observation was made for MP2, which uses silence insertion as PLC. Whether other PLC strategies may make bursty loss worse than random loss, as, e.g., in (Raake 2006), will be discussed by analysing the respective P.NAMS results for other codecs than the MP2 codec.

In Figs. 4.10, 4.11, and 4.12, the perceived quality is plotted for the P.NAMS database set as a function of the percentage of audio frame loss for different bitrates and burst lengths (the same legend convention as for MP2 was used), and for respectively AAC-LC, AC-3, and HE-AAC. Again, only one error pattern was used per combination of content and condition. The tendency that higher bitrate yields higher quality than medium and low bitrates in the case of losses is confirmed for all codecs. This observation is not as clear for AC-3 as for the other codecs, because both bitrates selected in the case of loss (96 and 256 kbps) yield high quality. For HE-AAC, isolated loss leads to higher perceived quality than bursty loss. This confirms that isolated loss may not be perceivable due to their short duration, and because any loss concealment works better with short-duration losses [see, e.g., (Raake 2006) for speech].

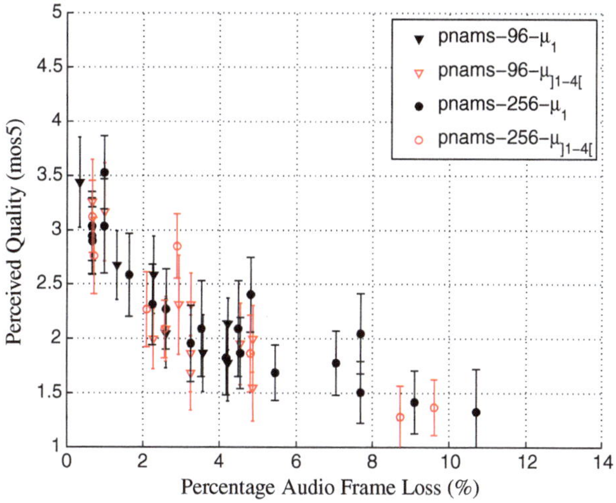

Fig. 4.11 Perceived audio quality for AC3 in the case of random and bursty losses, and for different bitrates (96, 256 kbps). μ: average burst length (see text for details). P.NAMS ($tr16$) database set

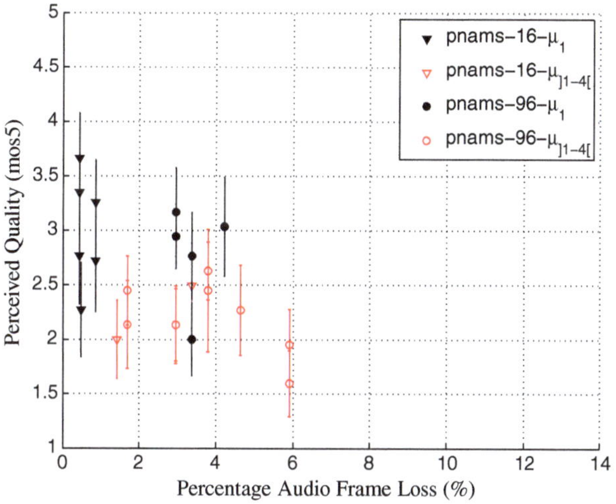

Fig. 4.12 Perceived audio quality for HE-AACv2 in the case of random and bursty losses, and for different bitrates (16, 96 kbps). μ: average burst length (see text for details). P.NAMS ($tr16$) database set

4.2 Parametric Audio Quality Model

4.2.1 State of the Art

Rix et al. (2006) provide an overview of existing audio and speech quality models. Both intrusive and non-intrusive methods, as well as signal- and parameter-based models are covered. In the context of this book, only the encrypted encoded bitstream, before the signal is decoded, can be accessed. The reference (non-degraded) and the decoded audio signal are not available. Moreover, for live monitoring, the complexity of the model should stay low. Therefore, the focus is on non-intrusive parametric audio quality models.

Most parametric models described in the literature (ITU-T Recommendation G.107 2005; Clark 2003; Broom 2006) are dedicated to speech communication links such as Voice over IP (VoIP). However, these approaches as well as the ones related to the parametrization of the effect of packet-loss in VoIP (Clark 2001; Raake 2006) may be used in the context of audio-links as well. For instance, Graubner et al. (2010) make use of the impairment-factor based approach of the E-model for predicting audio quality of Broadcast Audio over IP with a full-reference model.

The only non-intrusive parametric audio quality model known to the author is the model proposed by Egi et al. (2010). This model is suitable for both low and high bitrate applications such as MobileTV and IPTV. As input parameters it takes the audio bitrate, the codec type, the sampling rate, the frame length, the packet-loss-frequency (the number of loss events), and the average IP packet burst-length (considering only IP packets containing audio). The codec type, sampling rate, and frame-length are pre-determined. This model shows high correlation performance with subjective test results. However, it has been trained for one codec only (AAC-LC). Moreover, as noted by the authors, the interaction between audio bitrate and packet-loss-rate is not considered.

4.2.2 Audio Model

The audio quality model follows the impairment-factor-based approach described in Sect. 3.5.1, and which assumes that perceptual impairments linked with certain kinds of degradations can be considered as additive on a perceptual rating scale. The quality scores are thus first transformed onto the perceptual scale as already described in Sect. 4.1.2: For the TVM databases, the ratings are first averaged over all subjects, then converted from the 11-point scale to the 5-point Mean Opinion Score (MOS) scale and then transformed to the 100-point model-scale (Q_{100}) using the conversion defined in ITU-T Recommendation G.107 (ITU-T Recommendation G.107 2005) and reproduced in Appendix I of this work; for the P.NAMS databases, the ratings are averaged over all subjects and directly converted from the MOS-scale to the Q_{100}-scale using the same ITU-T G.107 transform.

Only TVM ($ph2_A$ and $ph3_A$) and P.NAMS training ($tr07$ and $tr16$) data were used for training the model, the P.NAMS validation set was used to validate the model. The audio quality model is then decomposed as follows:

$$Q_A = Qo_A - Icod_A - Itra_A \tag{4.1}$$

where Q_A is the predicted audio-quality and Qo_A is the base quality level the transmitted audio signal can reach for the respective target service. In this chapter, Qo_A is set to the transformed maximum quality rating which can be obtained in the audio subjective tests, i.e., $Qo_A = 100$. $Icod_A$ is the quality impact due to audio compression, and $Itra_A$ is the quality impact due to audio frame loss. $Icod_A$ is derived from the subjective tests for error-free conditions as follows: $Icod_A = Qo_A - Q_A$. Using all conditions, $Itra_A$ is obtained by computing $Itra_A = Qo_A - Icod_A - Q_A$.

The next step consists in selecting the best parameters for estimating $Icod_A$ and $Itra_A$, with the constraint that these parameters should be available from an encrypted audio stream, and therefore computable from packet-header information only. It can be observed from the analysis of the subjective test results that the bitrate and codec are good candidates for estimating $Icod_A$, and the audio codec, bitrate, percentage of audio frame loss, and burstiness (here, average number of consecutively lost audio frames) are good candidates for modeling $Itra_A$. Unfortunately, no audio-content-related parameter can be extracted from packet-header information for capturing the quality impact due to the audio content. The model thus provides the same estimate of quality for all contents.

In a least-square curve fitting procedure using the audio test results described in Sect. 4.1.2 as target values, the following equation is obtained for the frame loss-free case:

$$Icod_A = a1 \cdot \exp(a2 \cdot bitrate) + a3, \tag{4.2}$$

where $a1$, $a2$, and $a3$ are curve-fitting coefficients that depend on the used codec, and *bitrate* is the audio bitrate in kbps. Coefficient values are provided in Table 4.3, Sect. 4.2.3.

Following the same least-square procedure, Eq. (4.3) is obtained in the case of frame loss:

$$Itra_A = (b0 - Icod_A) \cdot \frac{Pfl}{(b1 \cdot \mu + b2) + Pfl} \tag{4.3}$$

Here, $b0$, $b1$ and $b2$ are the curve-fitting coefficients. The parameters Pfl is the percentage of lost audio frames and is expressed as follows:

$$Ppl = \frac{100 \cdot Total\ Number\ Of\ Lost\ Audio\ Frames}{Total\ Number\ Of\ Audio\ Frames} \tag{4.4}$$

μ captures the audio loss burstiness and is the averaged number of consecutively lost audio frames:

$$\mu = \frac{\sum_{k=1}^{N} Loss\ Event\ Array(k)}{N} \tag{4.5}$$

where N is the total number of loss events, k is the index of each loss event, and *Loss Event Array* denotes an array containing the number of consecutively lost audio frames for each loss event. As an example, if there are seven loss events, with $Loss\ Event\ Array = [2, 2, 1, 4, 2, 6, 1]$, then $\mu = \frac{2+2+1+4+2+6+1}{7} = 2.57$.

4.2.3 Model Evaluation

As a first step, a cross-validation (CV) procedure is run on the TVM ($ph2_A$ and $ph3_A$) and P.NAMS training ($tr07$ and $tr16$) databases for evaluating the parametric audio quality model and checking that the model has not been overtrained (see Sect. 2.4.3). As a second step, the model is validated against unknown data by using the P.NAMS validation databases $vl11$ and $vl12$. Note that the CV was needed only at that time the P.NAMS validation set was not available. Once the validation set is available, the CV is not required anymore, but it is noteworthy to report the results of the CV as an example of model validation when a limited number of databases is available.

4.2.3.1 Cross-Validation

Since the model will eventually be used for estimating the perceived audio quality of unknown contents, the cross-validation is done at source content level. For each iteration of the cross-validation, approximately 75 % of the contents are used for training the model (the CV training set), and the remaining 25 % are used for validating it (the CV validation set). There are as many CV iterations as permutations of the contents, in our case 28 permutations (see below). Contents are numbered from 1 to 8 in both P.NAMS training databases, and from 1 to 5 in the TVM database sets (see Tables 3.4 and 3.13). Contents 1–4 are similar for the two P.NAMS training databases, and contents number 1–5 are identical between the two TVM databases. As a consequence, content 5 actually corresponds to 3 different contents, and all other contents to 2 contents each, and there are 17 contents in total. As a consequence, when content 5 is part of the training set, the model is trained on more contents than when content 5 is not part of the training set. However, the deviation of the number of contents in the training set, whether content 5 is part of it or not, is quite low. For instance, if the model is trained on contents 1–6 and validated on contents 7 and 8, the model is actually trained on 13 contents (76 % of the contents) and validated on 4 contents (24 % of the contents), while when the model is trained on contents 1, 2, 3, 4, 6, and 7, and validated on contents 5 and 8, it is actually trained on

Table 4.1 Regression coefficients for AAC-LC resulting from the *CV*

Measure	a_1	a_2	a_3	b_0	b_1	b_2
AVG	100.00	−0.06	17.65	101.02	0.10	4.26
STD	0.00	0.00	0.97	1.33	0.00	0.28
MIN	100.00	−0.06	16.20	100.00	0.10	3.81
MAX	100.00	−0.05	19.79	104.21	0.10	4.78

AVG average over all CV runs, *STD* standard deviation, *MIN* minimum, *MAX* maximum

Table 4.2 Cross-validation performance results

Measure	R	RMSE	RMSE*
AVG	0.91	0.40	0.21
STD	0.01	0.02	0.03
MIN	0.89	0.37	0.16
MAX	0.93	0.45	0.26

12 contents (70 % of the contents) and validated on 5 contents (30 % of the contents). This content numbering system maintains a low deviation of the proportion of contents in the CV training and CV validation sets, while allowing contents from each database set to be kept in the CV training set, and thus avoiding any database bias effect. With this numbering system, 6 out of 8 contents (e.g., content numbers 1 to 6) are used for the CV training, and 2 out of 8 contents (e.g., contents 7 and 8) are used for the CV validation of the model. The contents are permutated for the CV training and CV validation sets, yielding 28 permutations, and thus 28 iterations of the CV.

At each permutation, the model is evaluated against subjective test results in terms of Pearson Correlation (*R*), Root-Mean-Squared-Error (*RMSE*, on the 5-point MOS scale), and epsilon-sensitive-*RMSE* (*RMSE**, also on the 5-point MOS scale) (see Sects. 3.5.3 and 2.4.3 for details).

A model is considered as overtrained when at least one of its coefficients take negative and positive values for different iterations of the CV. In that case, the CV is re-run on the model without the parameter corresponding to the most unstable coefficient, and performance indicators are re-computed.

A “large” variation of the Pearson correlation coefficients and RMSE over the different iterations is also an indication of instability of the model. How “small” the variation should be is let to the decision of the modeler.

The average (AVG), standard deviation (STD), minimum (MIN), and maximum (MAX) values of the resulting coefficients for the AAC-LC codec are reported in Table 4.1 as an example representative of all audio codecs.

The average, standard deviation, minimum and maximum values of the performance indicators are reported in Table 4.2.

The model shows high performance results. In particular, the predicted quality ratings are highly correlated with the subjective ones at each iteration of the cross-

Table 4.3 Regression coefficients per audio codec of the audio quality model

Codec	a_1	a_2	a_3	b_0	b_1	b_2
AAC	100.00	−0.06	17.65	100.26	0.10	4.18
AACc	100.00	−0.06	17.65	117.21	12.00	12.00
HE-AACv2	180.00	−0.14	25.07	115.71	0.10	6.60
HE-AACc	180.00	−0.14	25.07	200.00	34.44	34.45
MP2	100.00	−0.02	17.96	100.00	2.50	0.48
MP3	100.00	−0.02	0.00	100.00	0.60	0.58
AC-3	180.00	−0.03	20.82	100.00	0.10	3.10

AAC-LC and HE-AACv2 codecs: Nero implementation, without concealment; AAC-LCc and HE-AACv2c codecs: Dolby implementation of AAC-LC and HE-AACv2 codecs, with loss concealment

validation, with a minimum Pearson correlation coefficient of $R = 0.89$ and an average Pearson correlation coefficient of $R = 0.91$. The RMSE stays low for each iteration of the cross-validation, with a maximal value of $RMSE = 0.45$ and an averaged value of $RMSE = 0.40$ on the 5-point MOS scale.

These results indicate that the model was not overtrained and achieves high performance results on unknown contents and for typical IPTV degradations such as compression artifacts and transmission errors for five different codec.

4.2.3.2 Validation

The model is now trained on all available data (both P.NAMS training—$tr07$ and $tr16$—and TVM—$ph2_A$ and $ph3_A$—database sets). This results in the coefficient values reported in Table 4.3 for each audio codec.

The model is evaluated on the P.NAMS validation database set ($vl11$ and $vl12$) per audio files, i.e on 470 audio files (10 audio files had to be removed from the validation set due to processing errors). The same performance indicators as for the cross-validation are computed per database. For each performance indicator, the per-database values are then averaged over all validation databases. Results are reported in Table 4.4. See Sect. 3.5.3 for additional information on the computation of the overall model performance.

As mentioned in Sect. 3.5.3, the quality scores predicted by the audio quality model were first transformed to the 5-point MOS scale. Then a first-order polynomial mapping was applied per validation database on the transformed output and using the subjective ratings (averaged over all subjects) of each validation database as target values.

As can be observed from Table 4.4 and Fig. 4.13, the model shows high performance results with a Pearson correlation coefficient of $R = 0.92$, a Root-Mean-Square-Error of $RMSE = 0.39$, and an epsilon-modified RMSE value of $RMSE^* = 0.18$ (see Sect. 2.4.3).

The model is best performing for AC3, while the performance drops for the HE-AAC codec. This drop of performance is also visible in Fig. 4.13, where some black

Table 4.4 Performance results on unknown data

Codec	R	RMSE	RMSE*
All	0.92	0.39	0.18
AAC	0.92	0.34	0.12
MP2	0.93	0.38	0.19
HE-AAC	0.87	0.56	0.32
AC-3	0.95	0.28	0.08

All: all P.NAMS codecs (AAC, HE-AAC, MP2, AC-3)

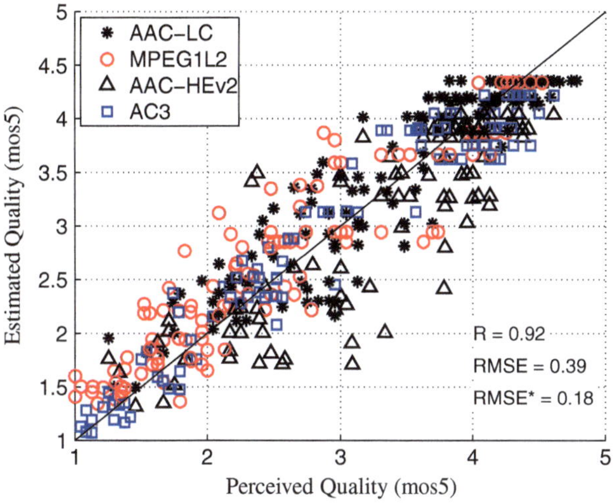

Fig. 4.13 Performance results on unknown data

triangles are further away from the correlation line. It is unfortunately still unclear why the performance drops in this case. No particular processing errors have been detected. However, the processing may have differed between $tr16$ and the validation databases, since the databases were produced by different laboratories. Looking again at Fig. 4.12, for which the plots are less smooth than for other codecs, the percentage of audio frame loss and the burstiness parameters μ do not capture the impact of audio loss as well as the other codecs. One possible explanation for this observation is a stronger influence of the content for that codec, and a stronger loss-location dependency and respective content interaction of the employed PLC.

The model shows saturation in the high quality range, with a maximum estimated audio quality of approx. $MOS = 4.4$. This behaviour is due to the fact that the model is not content-based and estimates the same audio quality for all contents.

Despite these drawbacks, the model achieves high performance results on unknown data. For comparison purposes, the audio module of the ITU-T P.1201.2 standard is implemented and performance results are computed on the P.NAMS validation database set. The same performance results as in Table 4.4 were obtained. The two audio quality models are therefore performing equally good on the P.NAMS validation set.

Note that for the purpose of the standardization, this audio module was ultimately trained on all available databases (TVM, P.NAMS training and validation databases). The P.NAMS validation set is thus known to the audio module of the ITU-T P.1201.2 standard.

4.3 Summary

Based on the results of audio quality perception tests, the quality impact due to audio compression and frame loss was analysed for five codecs (MPEG-I Layer II and III, MPEG-2 AAC LC, MPEG-4 HE-AACv2, AC-3), different codec implementations (for AAC-LC and HE-AAC) and thirteen contents typical of IPTV services. The HE-AAC and AAC codecs are the best performing codecs both in presence and in absence of frame loss. It was further observed that the behavior of the four codecs differs between speech and non-speech signals, especially for the Dolby implementation of HE-AAC and AAC. Finally, there is a tendency that long bursty losses ($\mu \geq 4$) yield better perceived quality than short bursty and random loss, while the loss location does not influence the perceived quality, except for HE-AAC. These analyses served as the basis for the development of a parametric audio quality model for IPTV services. This model takes as inputs the audio codec, the audio bitrate, the percentage of audio frame loss, and the averaged number of consecutively lost audio frames. Its output is an estimate of the perceived audio quality on the 100-point model-scale, which can be transformed onto the 5-point MOS scale. The model predictions show high correlation ($R = 0.92$) and low $RMSE$ ($RMSE = 0.39$, on the 5-point MOS scale) with unknown subjective test results. This model is similar to, and performs as well as the audio module of the ITU-T P.1201.2 standard on unknown data [see also (Garcia et al. 2013a, b)].

Chapter 5
Video Quality Model

This chapter presents two parametric video quality models: one model suitable for network planning, the other model applicable for quality monitoring and encrypted data. The network planning model is an extension of the model presented in Garcia and Raake (2010). The quality monitoring model is extending the model described in (Garcia and Raake 2011), which serves as baseline for the quality monitoring model. The subjective tests conducted for developing the models have already been introduced in Sect. 3.3. Test results, especially the quality impact due to the main IPTV degradations, compression and packet loss, will be analysed in Sect. 5.1. As mentioned in Chap. 3, the model type determines the model input parameters. Indeed, in the case of network planning, no measurement is done, and model inputs are only assumptions on the video transmission elements, such as the video codec or expected packet loss rate. The network planning model is presented and evaluated in Sect. 5.2. In the case of quality monitoring, more detailed information such as the type and size of a given frame can be extracted from the bitstream, allowing the computation of more accurate parameters, such as the spatial extent and the duration of loss-related video degradations. These parameters will be described in Sect. 5.3, and inserted into the quality monitoring model in Sect. 5.4.

As for the audio quality model, due to their more systematic modeling-directed approach, subjective test results from the TVM database serve as basis for the data analysis. The observations made on the TVM database set will be verified and extended on the P.NAMS training database set. The P.NAMS data will also be used for studying which frame-layer parameters, derived from the frame type and size information, can be used as input to the quality monitoring model.

Since the models were used for the ITU P.NAMS competitions, they were optimized on the P.NAMS training database set and evaluated following the P.NAMS validation procedure. As a first step, and because the P.NAMS validation set was not available during model development, a cross-validation procedure is applied on the P.NAMS training set to check that the model is not overtrained. The model is then validated against unknown data using the P.NAMS validation database set, once available.

M.-N. Garcia, *Parametric Packet-Based Audiovisual Quality Model for IPTV Services*, T-Labs Series in Telecommunication Services, DOI: 10.1007/978-3-319-04855-0_5,

The network planning video quality model—also referred to as "packet-based" model—on unknown data obtains a Pearson correlation coefficient of 0.84 with subjective ratings, and an RMSE of 0.56. The quality monitoring video quality model—also called "frame-based model"—outperforms the packet-based model on the same unknown data, with a Pearson correlation of 0.91 and an RMSE of 0.44. The frame-based model has been submitted as video quality module of the P.NAMS competition and obtained the highest performance results. It is now part of the ITU-T P.1201.2 standard.

5.1 Subjective Quality Assessment

5.1.1 Subjective Test Overview

As presented in Sect. 3.3, the video quality tests were conducted for collecting opinion ratings using an ACR test method with the continuous 11-point scale (ITU-T Recommendation P. 910 1999) in the TVM database sets, and with the 5-point categorical scale in the P.NAMS training and validation database sets. Twelve video quality tests were conducted in the TVM set, five in the P.NAMS training set, and eight in the P.NAMS validation set. A single video resolution was used per test. As described in Sects. 3.3.2 and 3.3.3, the same anchors (Tables 3.7 and 3.21) were used in all subjective tests. These anchors cover the whole quality range and degradation types used in the tests. They are used for mapping the results of the different tests, as described in Sect. 3.4.2.

The video contents used in the tests for the TVM, P.NAMS training and P.NAMS validation data sets are listed in Sects. 3.3.2 and 3.3.3, Tables 3.4, 3.13 and 3.14 respectively.

In order to simulate typical IPTV degradations, each video source sequence was processed off-line following the test conditions described in Table 3.5 for the twelve TVM databases, and in Tables 3.16 and 3.19 for the five training and eight validation P.NAMS databases.

For the TVM databases, this results in 228 video test conditions for each of the five video contents, leading to 1,140 video sequences to be rated by the subjects. For each of the P.NAMS databases, 30 test conditions were processed, resulting in 240 video files per P.NAMS database, thus 1,200 video files in total for the P.NAMS training set, and 1,920 video files for the P.NAMS validation set. Note that some video files were excluded from the TVM and P.NAMS training set due to processing error issues.

More information on the test designs and processing chains was given in the previous chapter, Sects. 3.3.2 and 3.3.3.

5.1.2 Subjective Test Results

The subjective test results analysis will be mainly conducted on the TVM database set and on the P.NAMS tr10 database.[1] The TVM set has the advantage of covering both H.264 and MPEG-2 codecs, but also to systematically test the quality impact due to the bitrate, the content, the GOP length (key-frame-rate), the loss burstiness, and the loss location. The P.NAMS tr10 will be analysed for extending the observed results to the 720p50 format, other slicing configurations and other contents. A detailed analysis of the content classification based on frame type and frame sizes will also be provided.

As for audio, the quality ratings of the TVM set were averaged over all subjects and error patterns (in the case of video packet loss and in order to study the influence of the loss location on the perceived quality, three error patterns were used per combination of content and condition, see Sect. 3.3.2.2). The obtained averages were first linearly transformed to the five-point MOS scale, then to the 100-point model-scale using the transformation defined in Appendix I of ITU-T Rec. G.107 (ITU-T Recommendation G.107 2005), and reported in the Annex I of this book.

For the P.NAMS subjective tests, quality scores were averaged over all subjects, then transformed from the five-point scale to the 100-point scale using the same G.107 transform. Only one error pattern was used per combination of content and condition in the case of packet loss.

Note that when results from a single database set are displayed, the original scale used in the tests (11-point scale for TVM, 5-point scale for P.NAMS) are used for the plots. This enables the use of confidence intervals as obtained in the tests. For TVM, results coming from the different databases have been merged. This means that the ratings of files coming from different databases, but corresponding to the same content and processed according to the same test condition, will be averaged. During the merging process, the confidence intervals were not calculated for the merged results and they are not available for this work. When results from the TVM and P.NAMS databases are merged, the 100-point model-scale is used in the graphs.

The analysis of the subjective test results is organized as follows: the first part is dedicated to the quality impact due to compression when no packet-loss occurs. It covers the quality impact due to bitrate, codec, key-frame-rate, and content. The second part analyses the quality impact due to packet loss for both slicing and freezing, and for both random and bursty losses.

5.1.2.1 Quality Impact Due to Video Compression

Figure 5.1a, b show the quality impact due to compression for H.264 and MPEG-2, for HD and SD, for the TVM database set, as a function of bitrate. Test results are averaged over all subjects and all contents. The influence of the content will be

[1] For copyright-related reasons, and with some exceptions, results from the P.NAMS databases produced by P.NAMS participants other than T-Labs cannot be published.

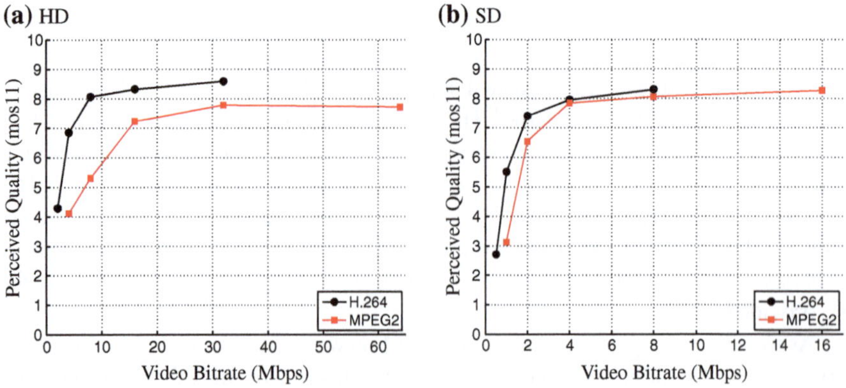

Fig. 5.1 Perceived video quality on the 11-point (*mos*11) scale as a function of the video bitrate for MPEG-2 and H.264. HD (**a**) and SD (**b**) resolution. TVM database set

analysed in the next section. Both curves follow the same exponential shape. As expected, for HD, H.264 performs clearly better than MPEG-2 at all bitrates. On average, the quality of H.264-encoded video is perceived as "good" (7 on the 11-point scale) already at 4 Mbps while a bitrate of 16 Mbps is necessary for MPEG-2 to obtain good video quality ratings. For SD, H.264 is significantly more efficient than MPEG-2 for low bitrates only. The quality of encoded SD video is perceived as "good" at 2 Mbps for H.264 and at 3 Mbps for MPEG-2. From 4 Mbps onwards, there is no difference of quality between the two codecs.

5.1.2.2 Quality Impact Due to Content

It is well known that the perceived quality of video is highly dependent on the video content, more specifically on the spatio-temporal complexity of the video.

In order to analyse the quality impact due to the video content for HD and SD, and for both H.264 and MPEG-2 coding, subjective ratings were averaged over all subjects per content and condition, i.e., per video file.

Figure 5.2a shows the quality impact due to compression as a function of the bitrate for the five contents used in the subjective tests in the HD case, when H.264 encoding is applied. It can be seen that the content has an influence on the perceived quality, especially at low and medium bitrates. Indeed, both at 2 and 4 Mbps, there is a MOS difference of two points (on the 11-point rating scale) between the best and worst perceived quality. From 8 Mbps onwards, this difference progressively decreases with the increase of the bitrate. At all bitrates, the soccer content (content C, see also Table 3.4) obtains the worst quality. This result is expected, since the soccer content has the highest spatial complexity ($SI \simeq 0.75$ in Fig. 3.3), due to the complex structure of the grass and of the crowd watching the soccer game. Interestingly, the movie content (content D) has a poor quality at low bitrate (2 Mbps), worse than

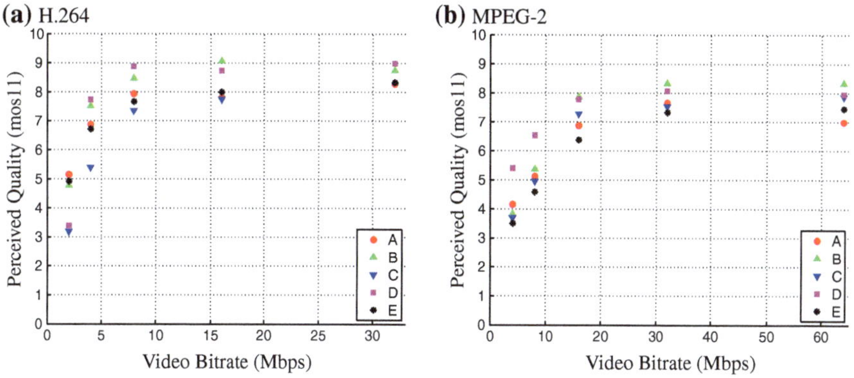

Fig. 5.2 Quality impact due to the bitrate per content for HD and for **a** H.264 **b** MPEG-2. TVM database set

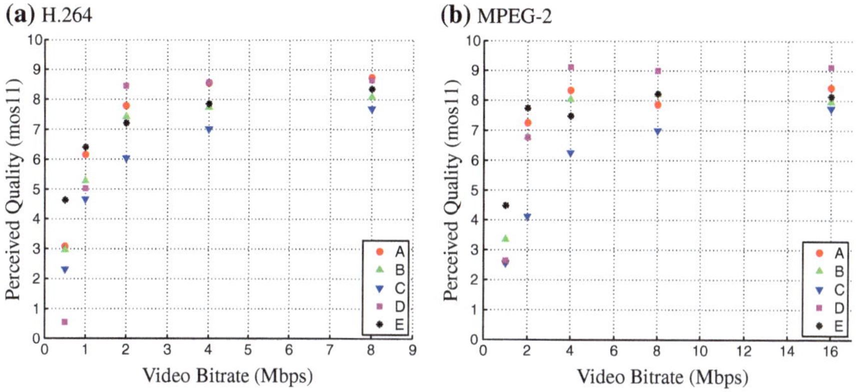

Fig. 5.3 Quality impact due to the bitrate per content for SD and for **a** H.264 **b** MPEG-2. TVM database set

the other contents except soccer, but obtains the best perceived quality for bitrates higher than 4 Mbps. This movie content has a high temporal complexity ($TI \simeq 80$ in Fig. 3.3) due to complex movements. It thus suffers from a too low bitrate, more than less temporally complex contents. However, it also contains black frames on top and bottom of the video. This characteristic could explain why it obtains the best perceived quality as long as the bitrate is sufficiently high for dealing with the high temporal complexity.

As can be seen in Fig. 5.2b, at low and medium bitrates, the movie content obtains better ratings than the other contents. It thus seems that the lowest bitrate used in the subjective tests is already sufficiently high for dealing with the high temporal complexity of the movie content. Note that the minimum bitrate used with MPEG-2 (4 Mbps) is higher than the minimum bitrate used with H.264 (1 Mbps), which

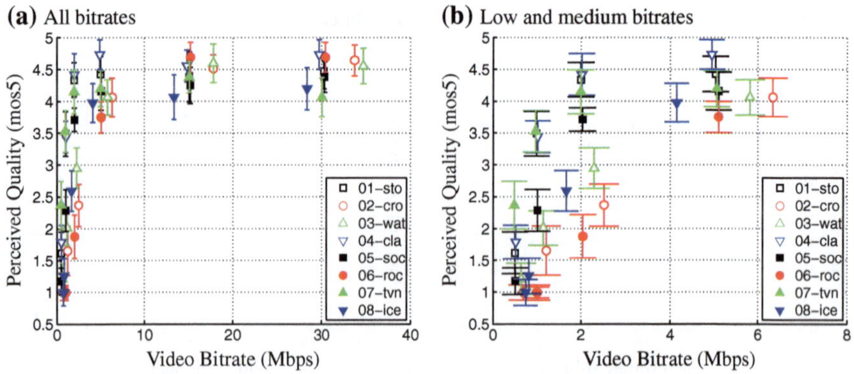

Fig. 5.4 Quality impact on the 5-point (*mos*5) scale due to the bitrate per content for HD720p, H.264, and **a** all bitrates, **b** low and medium bitrates only (P.NAMS, *tr*10)

may explain why none of the tested bitrates yields really poor quality for the movie content, instead of what was observed with H.264 in Fig. 5.2a.

It can be observed in Fig. 5.3a, b that similar results are obtained with SD.

The content dependency was also extensively analysed in the P.NAMS training databases, and conclusions drawn from the TVM data were confirmed on the P.NAMS data.

Figure 5.4a shows the perceived quality as a function of the bitrate for the P.NAMS database *tr*10 (HD720, 50 fps) and for the eight contents used in this database (see Table 3.13 for a more detailed description of the content). Results were averaged over all subjects per video file. As for TVM, the content dependency is at its strongest point for low and medium bitrates. For the sake of clarity, the small-to-medium bitrates range of Fig. 5.4a is reproduced in Fig. 5.4b. It can be observed that the source content "07-tvn" yields the highest quality at low bitrate. This content corresponds to TV news and is almost static ($TI \simeq 3$ in Fig. 3.5). Only the mouth of the TV speaker and sometimes his arm are moving. The temporal complexity of this content is thus really low and does not require many bits for being encoded. In contrast, source contents "02-cro" and "06-roc" are spatially and temporally much more complex than "07-tvn" ("02-cro":$SI \simeq 84$, $TI \simeq 75$; "06-roc": $SI \simeq 78$, $TI \simeq 33$ in Fig. 3.5). In particular, "02-cro" contains a scene in which many persons are running ("cro" stands for "cross-country"), and there are a lot of details, flash lights, smoke, and movements in content "06-roc". Due to their complexity, these contents obtain low quality scores at low and medium bitrates.

Section 5.3.1 analyses how these different content complexity levels can be derived from the video frame types (I, P, reference and non-reference B-frames) and sizes.

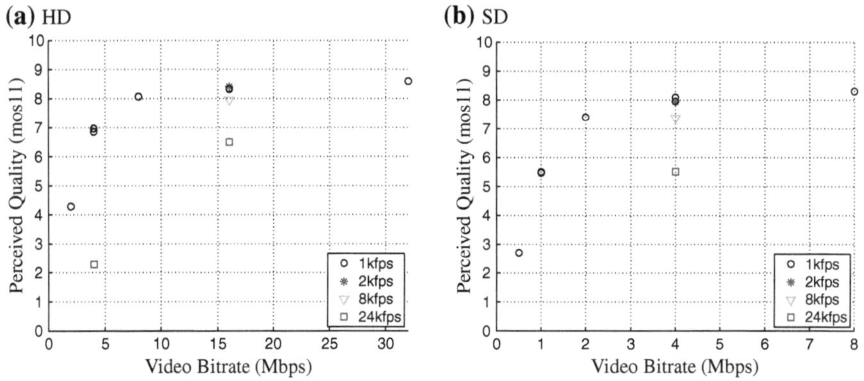

Fig. 5.5 Influence of the key-frame-rate on the quality impact due to the bitrate for H.264, and for **a** HD and **b** SD. *HD* 16 Mbps; *SD* 4 Mbps; kfps key-frame-per-second. TVM database set

5.1.2.3 Quality Impact Due to Key-Frame-Rate

In the context of MPEG-based video services, one of the parameters influencing the video perceived quality is the GOP-length, i.e., the distance between "I-frames" (see also Sects. 2.2.1 and 2.2.1.2). The GOP-length is generally chosen as a trade-off between encoding efficiency and error-propagation (Wu et al. 2006; Huszák and Imre 2010; Tektronix 1998). The key-frame-rate refers to the number of I-frames per second. As mentioned in Sect. 2.2.1.2, I-frames are intra-coded and are thus not predicted from previous frames. This is advantageous in the case of packet-loss and closed-GOP, since the loss is not propagated beyond a non-erroneous I-frame. However, since they are intra-coded, I-frames require more bytes than P- and B-frames for being coded and transmitted. Using too many I-frames per second may therefore lower the quality.

This behaviour can be observed in Fig. 5.5a, b. These figures show, for HD and SD, the influence of the key-frame-rate on the quality impact due to the bitrate. For both SD and HD, at high bitrates (16 Mbps for HD, 4 Mbps for SD), 1 and 2 kfps yield the same perceived quality. However, from 2 kfps onwards, the quality linearly decreases with the key-frame-rate. The linearity is more visible in Fig. 5.6a, b, which display the perceived quality as a function of the key-frame-rate for high and low bitrates (HD only). Note that for both SD and HD, the perceived quality of a video encoded at high bitrate (SD: 4 Mbps, HD: 16 Mbps) and with only I-frames ("24 kfps" in Fig. 5.5a, b) is equivalent to the perceived quality of a video encoded at one I-frame per second, but with a bitrate four times lower (4 Mbps for HD, 1 Mbps for SD, "1 kfps" in Fig. 5.5a, b). Obviously, a trade-off needs to be found between the robustness of the transmission system to packet loss and coding efficiency.

Finally, and as expected, the quality impact due to the key-frame-rate is bitrate-dependent and increases with a decreasing bitrate. This is reflected in Fig. 5.6a (HD) by the steeper slope of the curve corresponding to the low bitrate (4 Mbps).

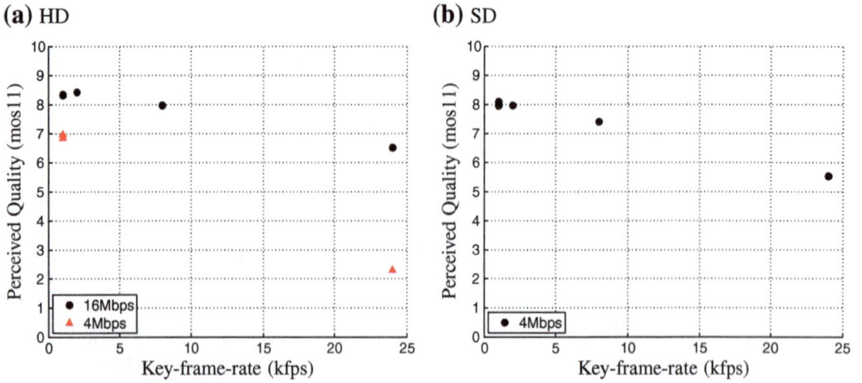

Fig. 5.6 Quality impact due to the key-frame-rate for H.264, and for **a** HD and **b** SD. kfps key-frame-per-second. TVM database set

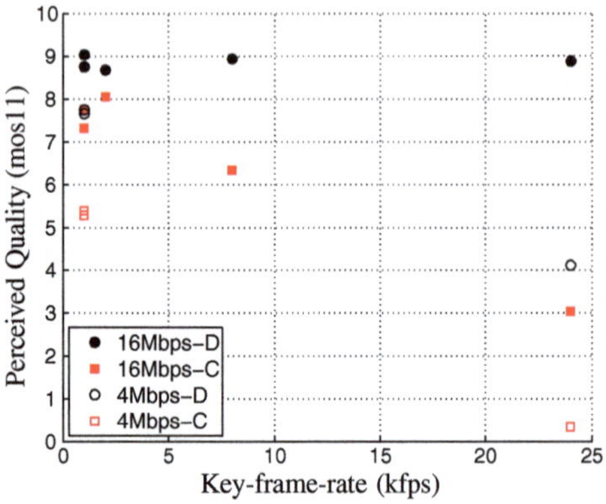

Fig. 5.7 Quality impact due to the key-frame-rate for HD, H.264, and for contents C and D. TVM database set

This effect depends, in addition, on the video content. Two extreme cases are shown in Fig. 5.7: At high bitrate (16 Mbps), the key-frame-rate has almost no influence on the perceived quality for the movie content (content *D*), yielding a flat curve. This is not the case for the soccer content (content *C*) for which the slope of the curve is steep. This result is in line with the observations made earlier in Sect. 5.1.2.2: Content D is of lower complexity, while content C and its high spatial complexity requires a lot of bits per frame.

For the P.NAMS database set, three types of GOP structures were mainly used: 1 key-frame per second (kfps) with non-hierarchical B-frames (IbbP), 2 kfps with non-hierarchical B-frames (IbbP), and 1 kfps with hierarchical B-frames (IbBbP). As can be seen in Fig. 5.8, the impact due to the key-frame rate is not significant when averaging the results over all contents. It should be noted that this result is in line with the results found on the TVM database, since no significant quality differences were observed for such small key-frame-rate variations (1 and 2 kfps).

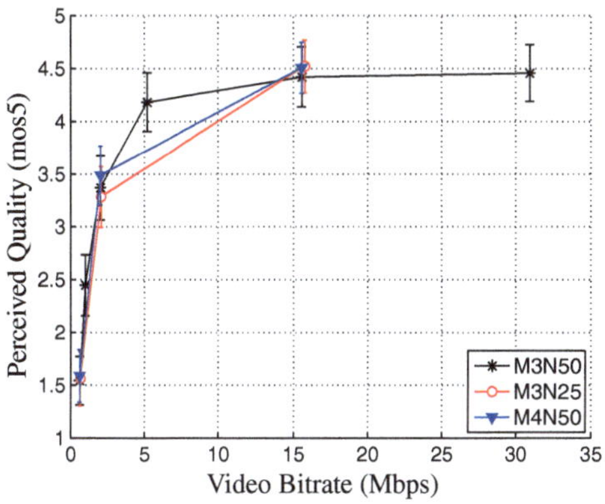

Fig. 5.8 Quality impact due to the bitrate for 720p50, H.264, and 3 GOP structures. *M3N50* 1 key-frame per second (kfps), IbbP; *M3N25* 2 kfps, IbbP; *M4N50* 1 kfps, IbBbP. P.NAMS, *tr10*

5.1.2.4 Quality Impact Due to Packet Loss

The elements influencing the perceived video quality, in the case of video packet loss, are listed below. All elements are not orthogonal to each others. The reader is referred to Sects. 2.2.1, 2.2.1.2 and 2.2.1.8, for a more detailed explanation of the video encoding and decoding process:

- The amount of lost packets,
- the packet loss pattern, which, in this work, mainly refers to the averaged number of consecutively lost packets,
- the GOP-structure, including:
  - the GOP-length,
  - the number and repartition of B- and P-frames in each GOP,
  - the GOP "type": Open-GOP versus closed-GOP.
- The frame type of the frame hit by loss. If the loss occurs on an I- or P-frame, the loss is propagated till the next I-frame; if the loss occurs on a B-frame, the loss is not propagated, except in the case of hierarchical coding, where loss in reference B-frames propagates to the dependent B-frames, i.e., till the next P- or I-frame.
- The number of packets per frame. The probability of hitting a frame with loss, and thus the probability of loss propagation if the hit frame is a reference frame, increases with the number of packets per frame. This number depends:
  - On the bitrate: When the bitrate increases, more IP packets are used for transmitting each frame.
  - On the frame type: Since they are intra-coded, I-frames require more packets than P- and B-frames for being transmitted. Being bidirectional, non-reference B-frames are encoded with less bits than P-frames, and therefore need less packets for being transmitted.

  - On the spatio-temporal complexity of the video: More packets are usually required for transmitting the I-frames in the case of high spatial complexity video; similarly, more packets are generally needed for transmitting P- and B-frames in the case of high temporal complexity video.

- The packet-loss-concealment: freezing versus slicing, with different strategies of concealing missing areas of affected slices.
- The number of slices per frame: The number of slices per frame is selected at the encoder stage. In the case of packet loss, and if slicing is applied, this amount influences the spatial extent of the loss. Indeed, if a packet loss occurs in a slice, the loss is propagated till the next slice, i.e., till the decoder can re-synchronize its decoding using the slice header. As a consequence, increasing the number of slices per frame reduces the spatial extent of the loss. However, this also increases the number of slice headers, and thus decreases the coding efficiency. A trade-off has to be found between coding efficiency and robustness to packet-loss.
- The spatio-temporal complexity of the content, which impacts the visibility of the loss. For instance, let's consider the case that the concealment consists in replacing the missing slice by the slice located at the same position in the previous frame. This concealment works well in the case of static or low motion content, such as TV news content, and the loss will not be visible. In the case of fast panning, this concealment is inefficient, and the loss is highly visible. In the case of spatially complex content, with a lof of spatial details, the concealment may not work well, but the visibility of loss is attenuated by the complex structure, in other terms, the loss is visually masked by the structure to some extents.
  The parametrization of these elements, and their integration into the model will be presented in Sect. 5.3.

5.1.2.5 Quality Impact Due to Error Pattern

In the TVM database, three different error patterns were applied for each combination of content and condition. This ensures varying the location of the loss for a given content, the amount of packet loss, and the burstiness.

The quality impact due to the error pattern can be observed in Fig. 5.9. These results highlight the need for more than one error pattern per packet-loss-rate when modeling the quality impact due to the packet-loss-rate without having access to the frame type. Indeed, and as mentioned in the previous section, the duration and thus quality impact due to packet-loss highly depends on the frame type in which the loss occurs.

In the case of the P.NAMS database set, only one error pattern was used per combination of content and condition, which prevents us from analysing the influence of the exact error pattern on the perceived quality for a given condition (in terms of packet-loss rate and distribution type).

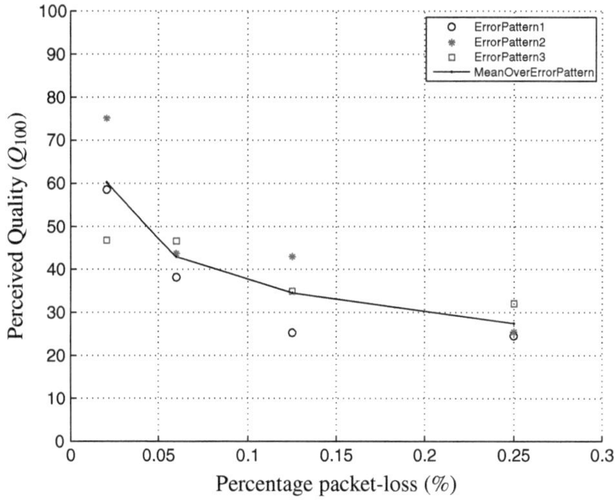

Fig. 5.9 Perceived quality of packet-loss, on the 100-point model scale, for three different error-patterns, and averaged over the three error patterns, in the case of freezing, and for content B (HD resolution TVM database set)

5.1.2.6 Quality Impact Due to Random (Uniform) Packet Loss

Figures 5.10a, b and 5.11a, b show, for HD and SD, the perceived quality of video sequences impaired by random loss at low (SD: 1 Mbps, HD: 4 Mbps), medium (SD: 2 Mbps, HD: 8 Mbps), and high (SD: 4 Mbps, HD: 16 Mbps) bitrates, for freezing and slicing, with one slice per Macroblock-row (MB-row). The quality ratings are averaged over all subjects, error patterns, and contents.

It can be observed that for both HD and SD, slicing yields much higher quality than freezing. For instance, for HD at high bitrate (16 Mbps), a perceived quality of $MOS = 3$ ("poor" on the 11-point rating scale) is obtained at a packet-loss-rate of 0.06 % for freezing and 0.5 % for slicing, that is a packet loss rate eight times higher in the case of slicing than in the case of freezing. This observation can be explained by the different types of distortion generated by slicing and freezing. Both distortions are time-varying. In the case of freezing, the video is jerky, while in the case of slicing and many slices per frame, the distortion does not disturb the smoothness/fluidity of the video. It is however expected that with the decrease of the number of slices per frame, the difference between freezing and slicing in terms of perceived quality will also decrease. For a low number of slices per frame (e.g., three slices per frame or even one slice per frame), slicing may even yield jerkiness (see Sect. 2.2.1 for a definition and description of "jerkiness"). For instance, if the loss occurs at the beginning of the slice, a wide proportion of the erroneous slice is replaced by the previous non-erroneous slice.

Figures 5.12b and 5.13b show the quality impact due to the packet-loss-rate in the case of slicing and for two encoder settings: for one slice per macroblock-row

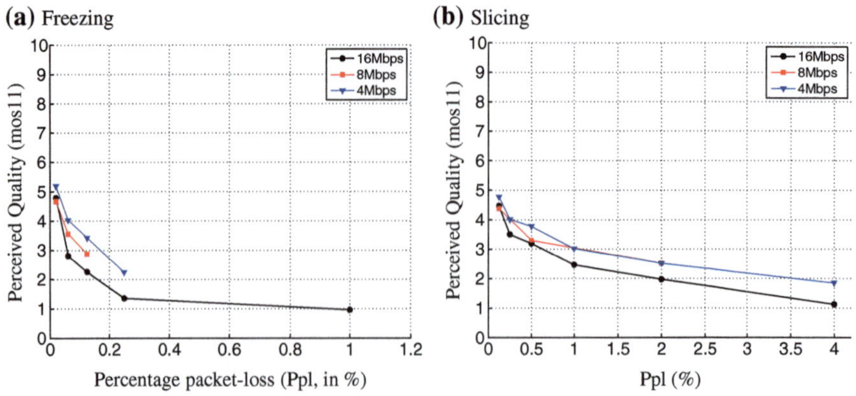

Fig. 5.10 Quality impact (on the 11-point *mos*11) due to random loss in the case of slicing and freezing (HD resolution, TVM database set). **a** Freezing, **b** Slicing

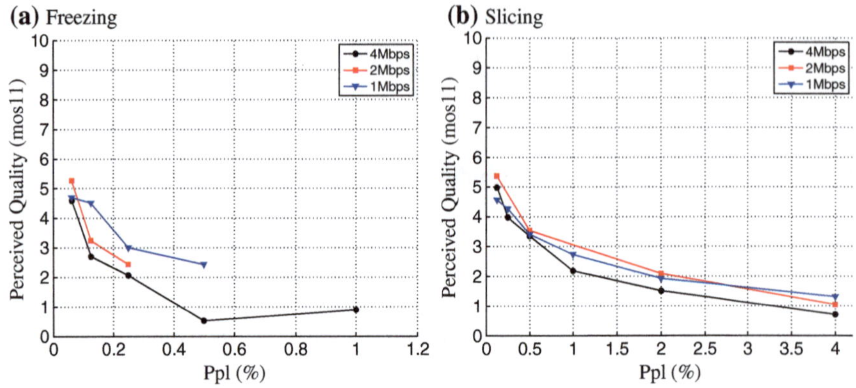

Fig. 5.11 Quality impact due to random loss in the case of slicing and freezing (SD resolution, TVM database set). **a** Freezing, **b** Slicing

and thus many slices per frame ("Nsl", $1088/16 = 68$ in the case of HD progressive video, $576/16 = 34$ in the case of SD progressive video), another one for three slices per frame ("3sl"). The freezing case is added in Figs. 5.12a and 5.13a for comparison. Note that for HD and high bitrates, freezing and three slices per frame are almost equivalent in terms of perceived quality, and in fact quality may even be lower ($Ppl = 1$ %, Fig. 5.12).

The quality decrease due to packet-loss in the case of three slices per frame is clearly stronger than in the case of 1 slice per Macroblock-line. This is expected, since, at least in the case of random loss, the spatial extent of the loss is higher in the three slices per frame setting. Indeed, and as previously mentioned, when packet loss hits a slice, the rest of the slice is also impaired, since the loss information cannot be recovered until the next slice header appears. In the case of random loss, single

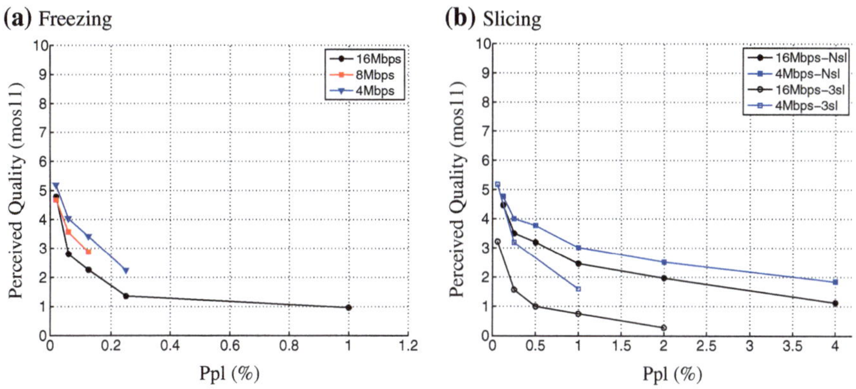

Fig. 5.12 Quality impact due to random loss in the case of slicing (**b**). Comparison with freezing (**a**). 1 slice per Macroblock-row (*Nsl*) versus 3 slices per frame (3*sl*) (HD resolution, TVM database set)

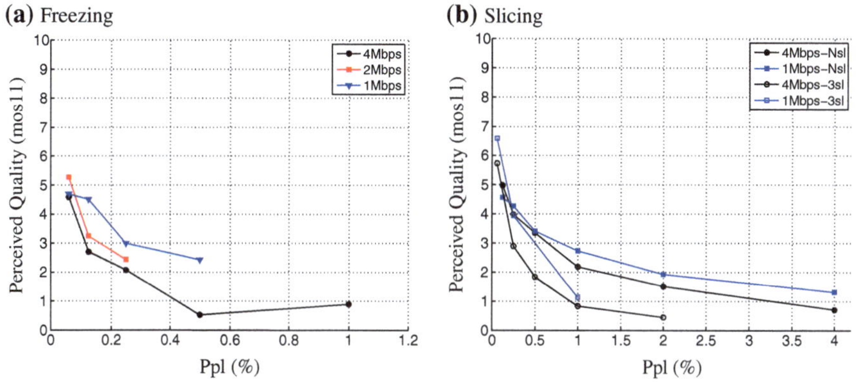

Fig. 5.13 Quality impact due to random loss in the case of slicing (**b**). Comparison with freezing (**a**). 1 slice per Macroblock-row (*Nsl*) versus 3 slices per frame (3*sl*). (SD resolution, TVM database set)

packets are lost most of the time. Since there are much less slices in the three-slices-per-frame configuration, one single lost packet results in a larger spatial extent of the loss than in the one-slice-per-macroblock-row case. This tendency is expected to decrease when the loss becomes more bursty. Indeed, if the loss is highly bursty, a high number of packets per frame are lost and the spatial extent of the loss is wide whatever the number of slices per frame is. This observation can partially explain contradictory results found in the literature on the quality impact due to bursty losses, the main explanation being probably the video resolution and the GOP structure.

Another interesting observation to be made on Figs. 5.12a and 5.13a is that in presence of packet-loss and in the case of freezing, the perceived quality is higher for low bitrates. Indeed, when the bitrate decreases, the number of packets per frame decreases, thus the probability of hitting a frame is lower, in particular the probability of hitting I- and P-frames, which contain more packets than B-frames, yielding shorter loss durations. Note that in the case of low bitrate, the quality of each non-distorted frame is lower than for high bitrate, in particular the quality of frozen frames are lower. This aspect does not seem to compensate for the longer loss duration due to high bitrate.

In the case of slicing (Figs. 5.10b, 5.11b, 5.12b and 5.13b), when the bitrate decreases and thus the number of packets per frame, each packet covers a larger area than with higher bitrates, yielding larger spatial extent of the loss. This compensates for the longer loss duration in the case of low bitrate, and explains why the bitrate seems to have a smaller influence in the case of slicing. Note that the quality difference between low and medium bitrates in the case of packet loss is not significant. This difference becomes significant between low and high bitrates, with lower quality in the case of high bitrates. This reflects that starting from a certain increase of bitrate, the loss duration becomes sufficiently longer, and that the quality impact due to longer loss duration is higher than the impact of larger spatial extent of the loss.

5.1.2.7 Quality Impact Due to Bursty Loss

Several studies have been conducted on the quality impact due to bursty packet losses. In Liang et al. (2003) and Lin et al. (2008), which address QCIF and CIF formats respectively, isolated losses are found to lead to better perceptual video quality than bursty losses. In Boulos et al. (2009), which addresses SD format, bursty losses are found to have generally better quality than isolated multiple losses. Those at first sight contradicting results show that the distribution of losses has a strong quality impact, but, following a deeper analysis of those results, that this impact depends on factors such as the video format, the packet-loss-concealment, the number of slices per frame, the GOP structure and the scene cut frequency.

Figures 5.14a, b and 5.15a, b show the quality impact due to bursty losses for slicing (one slice per MB-row) and freezing for HD and SD. For both PLC types, but in particular for freezing, bursty losses yield better quality than randomly distributed losses. Indeed, and in the case case of slicing, bursty losses yield larger extent of the loss, compared to random loss, but this is compensated by a smaller number of loss events. In the case of freezing, bursty losses do not necessarily yield stronger degradations, since it is equivalent loosing one or many packets in a given video frame, but they yield a smaller number of loss events—and thus better perceived quality—than in the case of random loss.

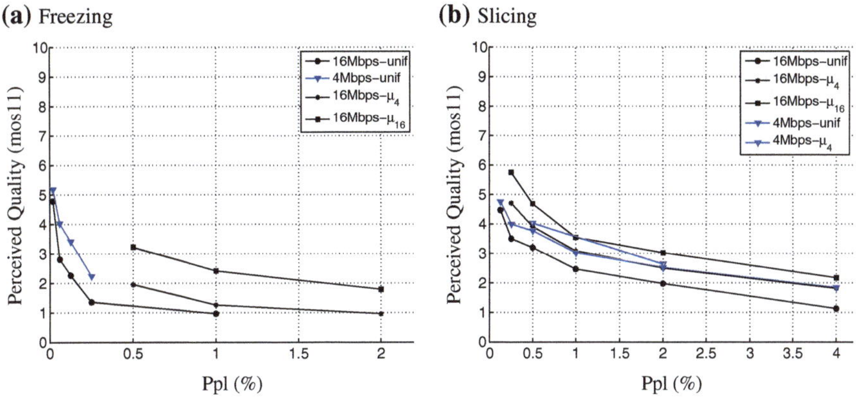

Fig. 5.14 Quality impact due to bitrate and loss distribution in the case of freezing and slicing; *unif* random (uniform) loss; μ_N: N packets lost in a row (HD resolution, TVM database set). **a** Freezing. **b** Slicing

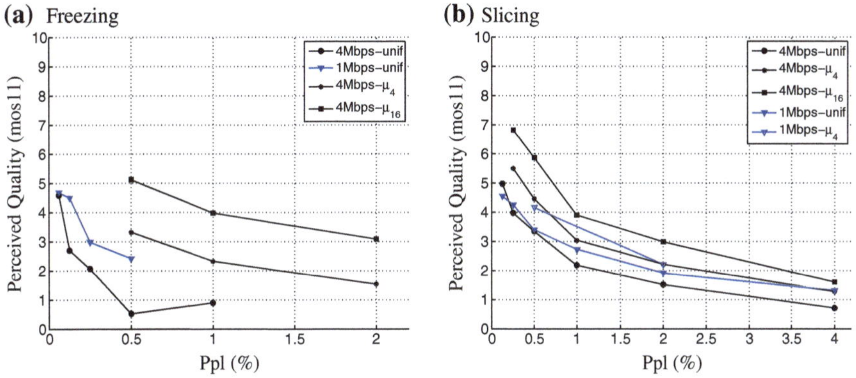

Fig. 5.15 Quality impact due to bitrate and loss distribution in the case of freezing and slicing; *unif* random (uniform) loss; μ_N: N packets lost in a row (SD resolution, TVM database set). **a** Freezing. **b** Slicing

5.2 Video Quality Model for Network Planning

5.2.1 State of the Art

In this section, a parametric packet-layer video quality model suitable for the network planning of IPTV services is presented. In the case of network planning, the signal is not available. Instead of network measurements, planning assumptions are made. Several studies are relevant to be mentioned in this context. In particular, Verscheure et al. in (1999) proposes a video quality model based on the average bitrate and packet-loss-rate, and Yamagashi et al. in (2008) developed a packet-layer model taking the average bitrate and packet-loss-frequency as inputs. Models found in the literature are generally not fully applicable to the scope of this work since they either address other applications and/or smaller formats [ITU-T Recommendation G.1070, (Tao et al. 2008)], random (uniform) loss only (Verscheure et al. 1999), or other types of error- and packet-loss-concealment (Yamagishi and Hayashi 2008).

In particular, Yamagashi et al. (2008) developed several packet-layer models, including one for video-telephony now specified in ITU-T Recommendation G.1070, and one for IPTV (Yamagishi and Hayashi 2008). The latter addresses HD resolution, but at 1,440 × 1,080 pixels (≡ 4 : 3 HD), and probably only for slicing (not specified in the chapter).

In Verscheure et al. (1999), Verscheure et al. propose a video quality model based on the average bitrate and packet-loss-rate. This model is not directly applicable in the present case, since it addresses variable bitrate, random loss only, and another type of error-concealment.

The model developed in Tao et al. (2008) covers both random and bursty losses, but for small formats only, while the focus of this book is on higher resolutions (SD, HD), and on both slicing and freezing.

Note that in Liang et al. (2003), conclusions are made based on the Mean Square Error (MSE) values and not on subjective test ratings. Finally, the authors in Lin et al. (2008) are addressing the SIF format only.

5.2.2 Packet-Based Video Model

The packet-based parametric video quality model developed for network planning follows the impairment-factor-based approach described in Sect. 3.5.1, which assumes that perceptual impairment linked with certain kinds of degradations are considered as additive on a perceptual rating scale. The quality scores are therefore first transformed onto the perceptual scale, and further transformed as the subjective test results analysis: for the TVM databases, the ratings are first averaged over all subjects, then converted from the 11-point scale to the 5-point Mean Opinion Score (MOS) scale, and then transformed to the 100-point model-scale (Q_{100}) using the conversion defined in (ITU-T Recommendation G.107 2005) (Annex I of this book); for the P.NAMS databases, the ratings are averaged over all subjects and directly converted from the MOS-scale to the 100-point scale using the same ITU-T G.107 transform. The video quality model is then decomposed as follows:

$$Q_V = Qo_V - Icod_V - Itra_V \tag{5.1}$$

where Q_V is the predicted video-quality and Qo_V is the base quality level the transmitted video signal can reach for the respective target service, and is set to the maximum value of the model scale ($Qo_V = 100$). $Icod_V$ is the quality impact due to video compression. It is derived from the subjective tests for error-free conditions as follows: $Icod_V = Qo_V - Q_V$. $Itra_V$ is the quality impact due to video packet loss. It is obtained using all conditions by computing $Itra_V = Qo_V - Icod_V - Q_V$.

The next step consists in selecting the best parameters for estimating $Icod_V$ and $Itra_V$, with the constraint that these parameters should be available during network planning. It can be observed from the analysis of the subjective tests that the video codec, the bitrate, and the GOP-length—or key-frame-rate—are good candidates for estimating $Icod_V$. In addition, for making the model independent of the video resolution and video frame rate, the number of bits per pixel *BitPerPixel* is used instead of the bitrate. The variable *Res* is further introduced for adjusting the impact of the video format.

The model of Eq. (5.2) is thus proposed for estimating $Icod_V$.

$$Icod_V = (a_1 * KeyFrameRate + a_4) \cdot exp(a_2 \cdot BitPerPixel \cdot Res) + a_3, \tag{5.2}$$

with $Res = 2$ for HD1080 and HD720 video, and $Res = 1$ for SD video, with

$$BitPerPixel = \frac{bitrate \cdot 10^6}{nx \cdot fr} \tag{5.3}$$

where *bitrate* is the video bitrate in Mbps, and *nx* and *fr* are the video resolution (number of pixels per frame) and the video frame rate, respectively.

a_1, a_2, and a_3 are obtained by applying a least-square curve fitting procedure using the transformed video test results described in Sect. 5.1.2 as target values. These curve-fitting coefficients depend on the used video codec. Coefficient values are provided in Table 5.3, Sect. 5.2.3.

In the case of packet loss, and based on the video test analysis of Sect. 5.1.2, the percentage of TS packet-loss, *Ppl*, and the average number of consecutively lost video TS packets, μ, are selected as parameters for estimating $Itra_V$. This results in Eq. (5.4):

$$Itra_V = (b_0 - Icodn_V) \cdot \frac{Ppl}{(Icodn_V \cdot (b_1 \cdot \mu + b_2) + Ppl)} \tag{5.4}$$

with

$$Icodn_V = 65 \text{ if } Icod_V > 65$$

and

$$Icodn_V = Icod_V \quad \text{otherwise.}$$

$Icod_V$ has been introduced in Eq. (5.4) to capture the interaction between the quality impact due to compression artifacts and transmission errors.

b_0, b_1, and b_2 are curve-fitting coefficients obtained by applying a least-square curve fitting procedure using the video test results described in Sect. 5.1.2 as target values. Coefficient values are provided in Table 5.3, Sect. 5.2.3. They depend on the applied packet-loss-concealment (freezing or slicing), and on the slicing configuration (one slice per frame vs. more than one slice per frame).

5.2.3 Model Evaluation

5.2.3.1 Cross-Validation

Similarly to the audio quality model, a cross-validation procedure is used for evaluating the parametric network planning video quality model. Since the model has been eventually developed within the P.NAMS competition context, and since the P.NAMS training database set contains a sufficiently high number of video-only databases (five), the cross-validation is done on the P.NAMS training database set only (*tr*08, *tr*09, *tr*10, *tr*11, and *tr*15). For ensuring that the whole quality range and all degradation types are present both in the CV-training and CV-validation sets, and since the model will eventually be used for estimating the perceived video quality of unknown contents, the cross-validation is conducted at source content level. For each iteration of the cross-validation, approximately 75 % of the contents is used for training the model, and the remaining 25 % is used for validating it. There are as many iterations as permutations of the contents, in our case 28 permutations (see below). For each iteration of the cross-validation, 6 contents are used for the training, and the remaining 2 contents are used for the validation, yielding 28 permutations.

At each permutation, the model is evaluated against the subjective test results in terms of Pearson Correlation Coefficients (R), Root-Mean-Squared-Error ($RMSE$, on the 5-point MOS scale), and epsilon-sensitive $RMSE^*$ (on the 5-point MOS scale as well). See Sect. 2.4.3 for a detailed description of these performance indicators.

As for audio, a model is considered as overtrained when at least one of its coefficients take negative and positive values for different iterations of the CV. In that case, the CV is rerun on the model without the parameter corresponding to the most unstable coefficient, and performance indicators are re-computed. A "large" variation of the Pearson correlation coefficients and RMSE over the different iterations is also an indication of instability of the model. How "small" the variation should be is let to the decision of the modeler, based on her/his experience.

Note that the variation of the key-frame-rate was small in the P.NAMS databases, in contrast to its variation in the TVM database. As a consequence, a first version of the model was developed on the TVM database to obtain values for the coefficient a_1. This value was used as lower boundary during the CV on the P.NAMS training databases.

Following this CV procedure, coefficient b_2 was forced to zero when more than one slice per frame was used as slicing configuration, i.e., $b_{2,Nsl} = 0$ in Table 5.1. The average (AVG), standard deviation (STD), minimum (MIN), and maximum (MAX) of the resulting coefficients are reported in Table 5.1. The corresponding average, standard deviation, minimum and maximum values of the performance indicators are shown in Table 5.2.

As can be observed in Tables 5.1 and 5.2, the standard deviation of the correlation coefficients and the RMSEs stay low, even though the proportion of contents in the training set varies between each iteration of the cross-validation. Moreover, the variation of the model coefficients values seem to stay low, at least they do not

Table 5.1 Cross-validation coefficients, using Eqs. (5.2) and (5.4)

	AVG	STD	MIN	MAX
a_1	10.00	0	10	10
a_2	−13.55	1.26	−16.70	−11.43
a_3	10.08	0.83	8.57	11.99
a_4	72.69	2.34	68.54	77.57
$b_{0,fr}$	68.86	0.70	67.50	70.27
$b_{1,fr}$	0.00018	0.00002	0.00015	0.00022
$b_{2,fr}$	0.00080	0.00018	0.00053	0.00116
$b_{0,1sl}$	80.72	2.45	76.49	85.46
$b_{1,1sl}$	0.00045	0.00006	0.00034	0.00056
$b_{2,1sl}$	0.00161	0.00091	0.00033	0.00357
$b_{0,Nsl}$	66.74	1.24	63.95	68.53
$b_{1,Nsl}$	0.00139	0.00016	0.00104	0.00166
$b_{2,Nsl}$	0	0	0	0

AVG average over all CV runs; *STD* Standard deviation; *MIN* Minimum; *MAX* Maximum $b_{x,fr}$, $b_{x,1sl}$, $b_{x,Nsl}$: regression coefficients obtained in the case of freezing, slicing with 1 slice per frame, and slicing with more than 1 slice per frame (i.e., 1 slice per MB-row) respectively

Table 5.2 Cross-validation performance results

	R	RMSE	RMSE*
AVG	0.84	0.59	0.40
STD	0.02	0.03	0.03
MIN	0.81	0.52	0.34
MAX	0.87	0.66	0.47

AVG average over all CV runs; *STD* Standard deviation; *MIN* Minimum; *MAX*: Maximum

present unwanted behavior like changing from negative to positive values from one iteration to another one. As a consequence, the model does not seem to be overtrained. Moreover, considering that the model is agnostic to the content, since it is a planning model, the model shows high performance results. In particular, the predicted quality ratings are highly correlated with the subjective ones at each round of the cross-validation, with a minimum Pearson correlation coefficient of $R = 0.81$ and an average Pearson correlation coefficient of $R = 0.84$. The *RMSE* stays low for each round of the cross-validation, with a maximal value $RMSE = 0.66$ and an averaged value of $RMSE = 0.59$ on the 5-point MOS scale.

These results indicate that the model achieves high performance results on unknown contents and typical IPTV degradations such as compression artifacts and transmission errors.

5.2.3.2 Validation on Unknown Data

The model is now trained on all P.NAMS training data (*tr*08, *tr*09, *tr*10, *tr*11, *tr*15, i.e., 1149 video files) and evaluated on the validation P.NAMS database set (*vl*13 to

Table 5.3 Regression coefficients of the network planning model, valid for all SD and HD formats

a_1	a_2	a_3	a_4
10.00	−13.42	10.05	72.40
$b_{0,fr}$	$b_{1,fr}$	$b_{2,fr}$	
69.39	0.00019	0.00082	
$b_{0,1sl}$	$b_{1,1sl}$	$b_{2,1sl}$	
80.61	0.00046	0.00147	
$b_{0,Nsl}$	$b_{1,Nsl}$	$b_{2,Nsl}$	
67.15	0.00144	0	

$b_{x,fr}$, $b_{x,1sl}$, $b_{x,Nsl}$: regression coefficients obtained in the case of freezing, slicing with 1 slice per frame, and slicing with 1 slice per MB-row, respectively

*vl*20, i.e., 1,920 video files). The training results in the coefficients shown in Table 5.3. As previously mentioned, and as can be observed in Table 5.3, the coefficient values depend on the applied packet-loss-concealment (freezing (*fr*) or slicing (*sl*)), and on the slicing configuration (1 slice per frame (1*sl*), and more than 1 slice per frame (*Nsl*), i.e 1 slice per Macroblock-row).

The same performance indicators as for the cross-validation are computed per validation database. For each performance indicator, the per-database values are then averaged over all validation databases. Results are reported in Table 5.4. See Sect. 3.5.3 for additional information on the computation of the overall model performance.

In addition, the estimated (predicted) quality scores are displayed as compared with the quality scores from the subjective tests in Fig. 5.16.

As mentioned in Sect. 3.5.3, the quality scores provided by the video quality model were first transformed to the 5-point MOS scale. Then, a first-order polynomial mapping was applied per validation database on the transformed output, using the subjective ratings (averaged over all subjects) of each validation database as target values.

As can be observed in both Fig. 5.16 and Table 5.4, for a network planning model the model shows high performance results, with a Pearson correlation coefficient of $R = 0.84$, a Root-Mean-Square-Error of $RMSE = 0.56$, and an epsilon-modified RMSE value of $RMSE^* = 0.37$.

It should be reminded that the network planning video quality model does not have access to the spatio-temporal complexity of the content. This explains why there is a "plateau effect" in Fig. 5.16: the model delivers the same estimate of quality for all contents, while subjective quality scores differ.

Table 5.4 further details the result per degradation type ("NoLoss", "Slicing" and "Freezing"). The "NoLoss" case achieves the highest correlation, but the worst (i.e., highest) *RMSE*. The Pearson Correlation is most probably misleading here. Indeed, and as can be observed in Fig. 5.16, the "NoLoss" case spans a quality range wider than the "Slicing" and "Freezing" cases. This artificially increases the Pearson Correlation value, without reflecting a better model prediction of the quality ratings.

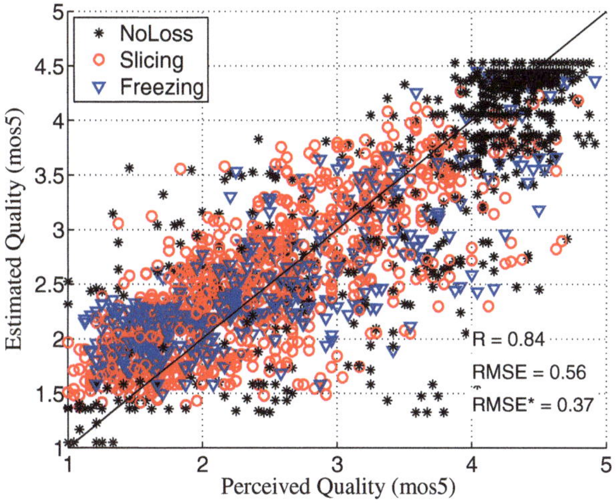

Fig. 5.16 Performance results of the network planning packet-based video quality model on unknown data. *NoLoss* No-loss cases only; *Slicing* slicing cases only; *Freezing* Freezing cases only

Table 5.4 Performance results. All: all degradation types (no packet loss, packet loss with slicing as PLC, packet loss with freezing as PLC)

Codec	R	RMSE	RMSE*
All	0.84	0.56	0.37
NoLoss	0.80	0.60	0.42
Slicing	0.68	0.54	0.33
Freezing	0.70	0.54	0.33

NoLoss performance results on video files without packet loss; *Slicing* performance results on slicing conditions; *Freezing* performance results on freezing conditions

The *RMSE* indicator here seems to be more appropriate for evaluating the model performance. In particular, it better reflects what can be observed from Fig. 5.16: Data points corresponding to the "No Loss" case are generally further away from the correlation line than for "Slicing" and "Freezing". This result may reflect that the content dependency, which is not captured by the model, is stronger in the "NoLoss" case. From both the figure and the *RMSE* values, it can also be concluded that the model achieves the same accuracy in case of "Slicing" and "Freezing". In any case, with a variation between 0.56 and 0.60, the *RMSE* values are relatively stable among the different degradation types.

There is however room for improving the model performance, especially for the case of quality monitoring, for which frame-related information can be extracted from the encrypted bitstream. This may help in capturing quality-relevant information not available in the network planning scenario, such as the content complexity, and estimates of the spatial and temporal location of the loss.

This is the scope of the next section, which provides a parametric description of the video content, slicing, and freezing degradations. In this case, parameters are derived from video-frame-related information and more accurately capture the quality impact due to compression artifacts and transmisson errors.

5.3 Perceptually Motivated Frame-Based Parameters

5.3.1 Parametric Description of Content

In the case of quality monitoring and encrypted data, the frame type is unknown. However, information about the frame boundaries can be extracted from the encrypted bitstream (see Sect. 2.2.1.6), and the frame size—here, the amount of bytes per frame—can be calculated. The frame types can then be estimated from the frame sizes and by making assumptions on the GOP structure. A general procedure for estimating the frame type is described in (List et al. 2012).

Once the frame sizes and types are known, further assumptions can be made about the content type, especially on the spatio-temporal complexity of the video.

In this section, the impact of the content when no packet loss occurs is investigated. The quality impact due to the content in the case of packet loss and in the case of slicing and freezing is investigated separately in the following sections.

Figures 5.17, 5.18 and 5.19 display the frame sizes of each frame of the video sequence as a function of the frame index for contents 6 ("06-*roc*"), 7 ("07-*tvn*"), and 8 ("08-*ice*") of the *tr*10 database (see Table 3.13), at low (1 Mbps) and high (15 Mbps) bitrates. For facilitating the comparison of frame sizes between contents, the y-axis of the plots are scaled to the same range for a given bitrate.

The content "07-*tvn*" has a low spatio-temporal complexity (Broadcasting News content, for which the camera is fixed and the presenter moves barely more than his mouth), and obtains the highest quality (between "poor" and "fair" in Fig. 5.4) for low bitrate, while "06-*roc*" and "08-*ice*" obtain for approximately the same low bitrate as "07-*tvn*" the lowest quality ("bad"). Content "08-*ice*" has, in addition, the advantage of having two scenes of different spatio-temporal complexity, which allows comparing the intra- and inter-scene variation of the frame sizes. It can be observed in Figs. 5.17, 5.18 and 5.19 that the frame sizes are not stable during the first GOP. This corresponds to the encoder stabilization phase which occurs in the case of single-path encoding. The first GOP, or at least the first I-frame should therefore be removed from the analysis and computation of content-related features. Once the first GOP of the sequence (or of a scene) has been excluded, the I-frame sizes and the ratio of the B- over P-frames do not vary much over a scene, especially when comparing this variation to the inter-scene variation.

The I-frame sizes seem to play an important role in the determination of the content complexity. In particular, Fig. 5.18 shows that the I-frame sizes are much bigger for content 7, which has a low spatio-temporal complexity. It is not intuitive

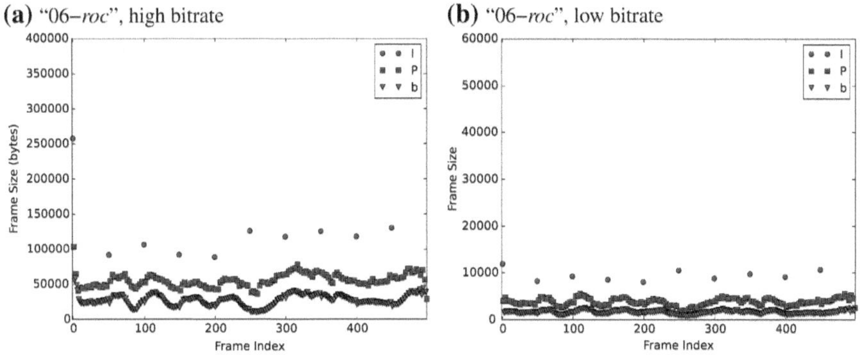

Fig. 5.17 Frame sizes as a function of the frame index for high and low bitrates. P.NAMS, *tr*10, content "06-*roc*" (cf. Table 3.13). **a** "06-*roc*", high bitrate. **b** "06-*roc*", low bitrate

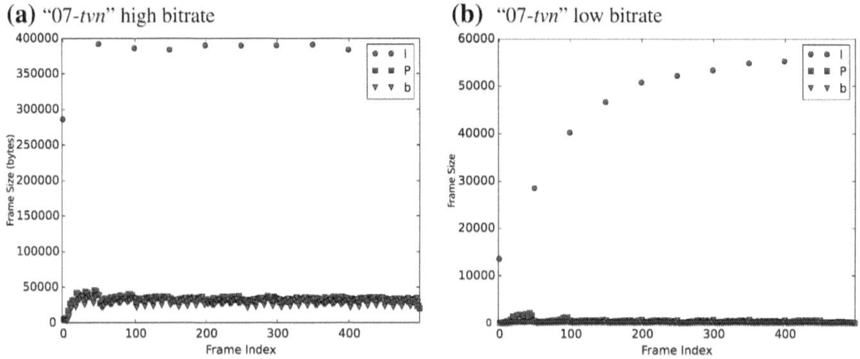

Fig. 5.18 Frame sizes as a function of the frame index for high and low bitrates. P.NAMS, *tr*10, content "07-tvn" (cf. Table 3.13). **a** "07-*tvn*" high bitrate. **b** "07-*tvn*" low bitrate

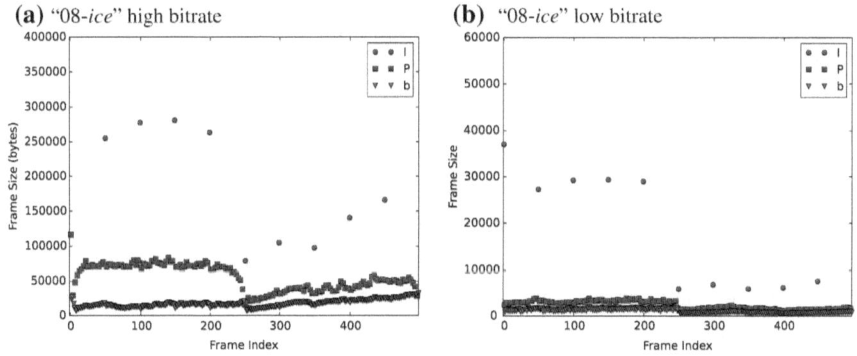

Fig. 5.19 Frame sizes as a function of the frame index for high and low bitrates. P.NAMS, *tr*10, content "08-*ice*" (cf. Table 3.13). **a** "08-*ice*" high bitrate. *b* "08-*ice*" low bitrate

since low I-frame sizes are expected for low spatial complexity content. However, this content has, in addition, a low temporal complexity, yielding low P- and B-frame sizes, and the bits seem to primarily be allocated to the I-frames.

The following *scene-complexity* features seem to be of interest for characterizing the video contents:

- S^I_{sc}: Average I-frame size for the given scene sc. The first I-frame of the first scene is ignored in the case of single-path encoding,
- S^P_{sc}: Average P-frame size for scene sc,
- S^B_{sc}: Average size of reference B-frames (used in the case of hierarchical coding) per scene sc,
- S^b_{sc}: Average size of non-reference B-frames—"b" frames—per scene,
- S^{noI}_{sc}: Averaged P-, B- and b-frame sizes per scene.

Note that a (video) scene starts with an I-frame and generally contains several GOPs. In the case of encrypted video, the scene cuts can be detected using (Argyropoulos et al. 2013). Two video scenes usually differ by their semantic contents. Moreover, the intra-scene variation of the spatio-temporal complexity of the content is generally lower than its inter-scene variation.

The following ratios are also considered as scene-complexity parameters. Each ratio is computed per scene from the scene-complexity parameters previously defined:

- $S^{P/I}$: S^P_{sc}/S^I_{sc},
- $S^{B/I}$: S^B_{sc}/S^I_{sc},
- $S^{b/I}$: S^b_{sc}/S^I_{sc},
- $S^{noI/I}$: S^{noI}_{sc}/S^I_{sc},
- $S^{b/P}$: S^b_{sc}/S^P_{sc}.

Figures 5.20a and 5.21a show the most relevant features (S^I_{sc} and $S^{P/I}_{sc}$), taken from the above list, as a function of the bitrate for the "M3N50" GOP structure (1 key-frame every 50 frames (i.e., per second), two B-frames between P-frames) in database $tr10$. It has been observed that a low averaged I-frame size at low bitrate seems to indicate low quality. Moreover, it can be expected that the scene with the lowest quality will have more influence on the overall perceived quality than the other scenes. For the sake of clarity, only the lowest S^I_{sc} values and the highest $S^{P/I}$ values of each sequence are indicated in Figs. 5.20a and 5.21a.

It can be observed that the S^I_{sc} and $S^{P/I}_{sc}$ features are well isolating content "07-*tvn*", which obtains the highest values for S^I_{sc} and the lowest values for $S^{P/I}_{sc}$. Compared to $S^{P/I}_{sc}$, the feature S^I_{sc} more consistently places the contents "02-*cro*" and "06-*roc*" at the opposite side of "07-*tvn*" in the graph. These results are to be compared to the perceived quality as a function of the bitrate for the different contents (Fig. 5.4b). For convenience, this figure is reproduced besides Figs. 5.20a and 5.21a. The observations made on S^I_{sc} are promising since for low bitrate, "07-*tvn*" obtains the highest quality while "02-*cro*" and "06-*roc*" lie on the lowest part of the quality curve. The only

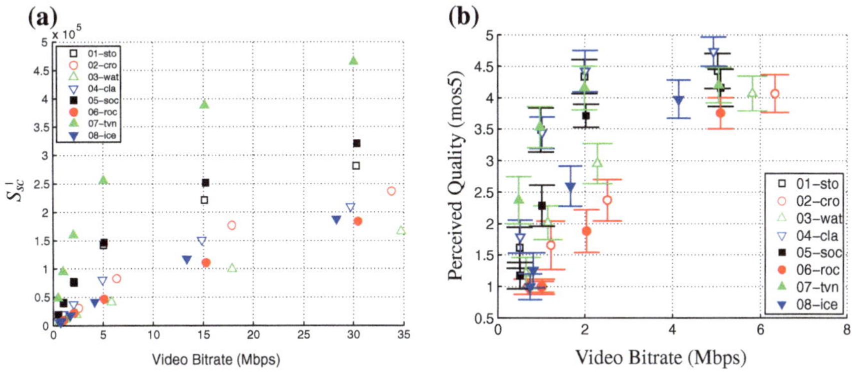

Fig. 5.20 Frame-size related parameter S^I_{sc} (**a**) and perceived quality (**b**) versus video bitrate. P.NAMS, *tr*10

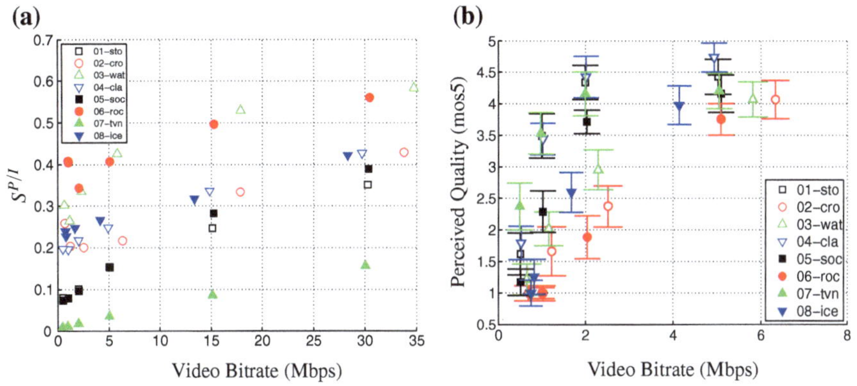

Fig. 5.21 Frame-size related parameter $S^{P/I}$ (**a**) and perceived quality (**b**) versus video bitrate. P.NAMS, *tr*10

content for which S^I_{sc} may produce outliers is "04-*cla*", which has low S^I_{sc} values but obtains similar perceived quality as "01-*sto*" and even "07-*tvn*" from 1 Mbps. "04-*cla*" contains a strong left-right panning but a lot of black areas as well. As a consequence, the panning may require higher P- and B/b-frame sizes, penalizing the I-frames by reducing by some bits, but probably without impacting the overall quality. Indeed, due to the large black area, the I-frames do not require many bits. From packet-header information, black area cannot be detected.

Despite this drawback, the S^I_{sc} feature is used as basis for predicting the content dependency in the case of coding. One more observation motivated this choice: The S^I_{sc} curves (Fig. 5.20a) follow a logarithmic shape, the S^P_{sc} curves rather a linear shape (see Fig. 5.22a) and the S^b_{sc} curves an exponential shape (see Fig. 5.22b). This indicates that the S^I_{sc} values are strongly increasing with the bitrate for low to medium bitrates, while the S^b_{sc} values are more strongly increasing from medium bitrates onwards. The bit rate allocation algorithm is thus favouring the I-frames first, then

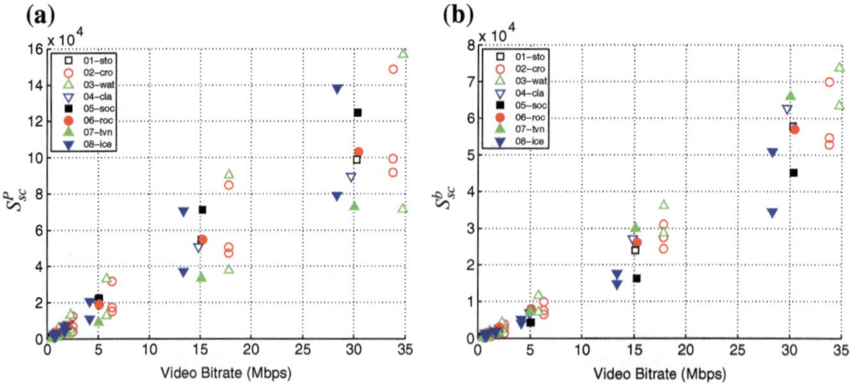

Fig. 5.22 Frame-size related parameters S^P_{sc} (**a**) and S^b_{sc} (**b**) versus video bitrate. P.NAMS, $tr10$

the P-frames, and lastly the (B- and) b-frames. As a consequence, the difference of I-frame sizes between contents at low bitrate is carrying most of the content-complexity-related information. This supports the use of the S^I_{sc} values for capturing the impact of the content at low bitrate.

It should be noted that the parameters derived from the above features and to be used in the model should not vary with the GOP structure, except if the perceived quality varies as well. Moreover, and as observed in Fig. 5.4, the content influences the perceived quality for low bitrate only. As a consequence, the selected features or parameters derived from the features should distinguish the contents for low bitrate only. In addition, there typically are several scenes in a sequence, and since the selected features are computed per scene, it has to be determined how the scene-based features should be combined into a sequence-based feature. Finally, the parameter should be independent of the frame rate and of the video resolution.

For all these reasons, the following *SceneComp* parameter is proposed:

$$SceneComp = \frac{\sum_{sc=1}^{Z} w_{sc} \cdot N_{sc}}{\sum_{sc=1}^{Z} S^I_{sc} \cdot w_{sc} \cdot N_{sc}} \cdot \frac{nx \cdot fr}{1{,}000}, \tag{5.5}$$

where S^I_{sc} is one of the *scene-complexity* feature described above, nx and fr are the video resolution (number of pixels per frame) and the video frame rate, Z is the number of scenes in the video sequence, and N_{sc} is the number of GOPs in scene sc. For the scene having the lowest S^I_{sc} value, $w_{sc} = 16$, otherwise $w_{sc} = 1$.

This parameter will be included in the video quality model in Sect. 5.4.2 and evaluated in Sect. 5.4.3.

5.3.2 Parametric Description of Slicing Degradation

In Sect. 5.1.2, it has been observed that the location of the loss influences the perceived quality. The packet loss rate and packet loss distribution are not sufficient for accurately describing the quality impact due to packet loss. As seen in Sect. 5.1.2, several additional elements are influencing the impact of packet loss on the perceived quality, such as the type of the frame hit by loss, the number of slices per frame, or the spatio-temporal complexity of the content. All these elements are eventually impacting the spatial extent, the duration, and the visibility of the loss. As a consequence, the existence of a generic parameter, *xwpSEQsz*, describing the spatial extent, the duration, and the visibility of the loss is investigated. With this parameter, it is assumed that all degradation events occurring in a short period of time—typically 10 s, the duration of the video shown to the test subjects—are perceived as a single overall degradation entity by the persons assessing the perceived video quality.

5.3.2.1 Spatial Extent and Duration of the Loss

xwpSEQsz is expressed as follows (see also Raake et al. 2012):

$$xwpSEQsz = \frac{\sum_{k=1}^{G} \alpha_k \cdot xl_k}{\sum T_k}, \tag{5.6}$$

with $xwpSEQsz \in [0, 1]$, and T_k being the GOP length for GOP k. α_k ($\alpha_k \in [0, 1]$) is a weighting factor depending on the spatio-temporal complexity of the content which will be described later in this section. G is the number of GOPs in the video sequence. xl_k is the spatio-temporal description—i.e., the magnitude—of the loss for GOP k and is expressed as follows:

$$xl_k = \sum_{i=1}^{L} xle_i \cdot (T_k - t_i), \tag{5.7}$$

t_i is the location in the GOP of the loss event i, $T_k - T_i$ is therefore the duration of loss event i, L is the number of loss events in GOP k, and xle_i is the spatial extent of the loss event i.

In the case of one slice per frame:

$$xle_i = \frac{nap_i}{np_i}, \tag{5.8}$$

Here, nap_i is the number of affected TS packets in the hit frame for the loss event i. It should be reminded that, when a packet is lost in a slice, the decoder cannot re-synchronize its decoding before the next slice header. As a consequence, in the case of one slice per frame, nap_i corresponds to the number of TS packets between the first lost TS packet in the frame and the last TS packet of the frame.

In the case of more than one slice per frame, since the number of packets per slice and the packet location in the slice are both unknown, the exact spatial extent of the loss cannot be computed from packet-header information. However, and as shown in Raake et al. (2012) and Frossard and Verscheure (2001), half-a-slice is lost on average and in addition to the proportion of impaired pixels corresponding to the loss event in the hit frame. As a consequence, the spatial extent of the loss in the case of more than one slice per frame is expressed as follows:

$$xle_i = \frac{nlp_i}{np_i} + \frac{nle}{2 \cdot ns}, \tag{5.9}$$

where nlp_i is the number of lost TS packets in the current hit frame (for the loss event i), np_i is the total (lost + found) number of TS packets in the current hit frame, nle is the number of independent loss events in the current hit frame; one independent loss event can contain one or more consecutive lost packets; two independent loss events are separated by at least one non-impaired slice, i.e none of the packets of this slice is lost. In the case of encrypted data, it is not known to which slice one packet belongs to. As a consequence, two loss events are said independent when they are separated by at least one found packet. ns is the number of slices per frame, $\frac{nlp_i}{np_i}$ is the proportion of impaired pixels in the current hit frame corresponding to the pixel-area encoded in the lost packets. $\frac{nle}{2 \cdot ns}$ is a correction term reflecting that when there is a loss event in a slice, on average half-a-slice is lost in addition to the proportion of impaired pixels corresponding to the number of lost packets in the hit frame (i.e., to $\frac{nlp}{np}$).

It should be noted that the size of the loss cannot extend the size of the frame. For taking into account that observation, let's define in Eq. (5.10) the cumulative spatial extent of all loss events up to the current loss event i in the GOP.

$$XLE_i = \sum_{j=1}^{i} xle_j \tag{5.10}$$

Then the computation of xle_i in Eqs. (5.8) and (5.9) if modified as follows:

In the case of one slice per frame:

$$xle_i = \begin{cases} \frac{nap_i}{np_i} & \text{if } XLE_i < 1 \\ 1 - XLE_{i-1} & \text{if } XLE_i \geq 1 \end{cases} \tag{5.11}$$

and in the case of more than one slice per frame:

$$xle_i = \begin{cases} \frac{nlp_i}{np_i} + \frac{nle}{2 \cdot ns} & \text{if } XLE_i < 1 \\ 1 - XLE_{i-1} & \text{if } XLE_i \geq 1 \end{cases} \tag{5.12}$$

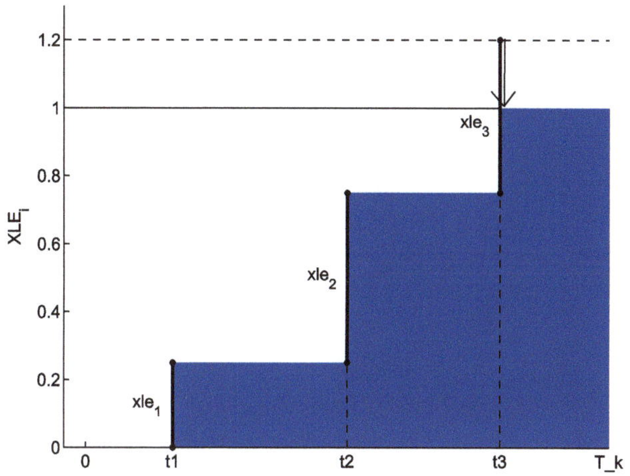

Fig. 5.23 Loss magnitude xl_k (*blue area*) in GOP k. Computation example for three loss events

Figure 5.23 shows an example of the computation of xl_k in the case of three loss events happening in GOP k at time t_1, t_2, and t_3. xl_k is represented by the filled blue area, and xle_1, xle_2, and xle_3 by the bold vertical dark segments. The cumulated spatial extent of the loss ($xle_1 + xle_2 + xle_3$) is bigger than 1. However, and as can be observed in the figure, the constraint defined in Eq. (5.10) and represented by the arrow in the figure, prevents the cumulated spatial extent of the loss for being bigger than 1.

Figure 5.24b shows the perceived quality as a function of *xwpSEQsz* for *tr*10 when no content-dependency is applied, i.e., when $\alpha_k = 1$ for all GOPs k in Eq. (5.6). For comparison, the perceived quality as a function of the "Percentage packet-loss", *Ppl*, is provided in Fig. 5.24a. In this example, both random and bursty packet losses are covered. The bitrates are varying between 5 and 15 Mbps. As a consequence, there is no compression artifact in the video sequences. The interaction between compression and slicing artifacts will be studied later in this section.

It can be observed in Fig. 5.24a, b that there is a clear relation between the *xwpSEQsz* parameter and the perceived quality, which is clearer than the relation between the quality and the packet-loss percentage.

5.3.2.2 Content Dependency

The influence of the content on the perceived quality is clearly visible in Fig. 5.24b, in which the TV news content "07-*tvn*" obtains much higher quality than the other contents for the same *xwpSEQsz* value. This behavior is expected since, as previously mentioned, the TV news content has a low temporal complexity (static camera, almost no movement in the scene). As a consequence, when a lost slice is replaced by the corresponding non-erroneous slice from the previous frame, the loss is barely visible. In contrast, for contents with high temporal complexity—or containing scenes with high

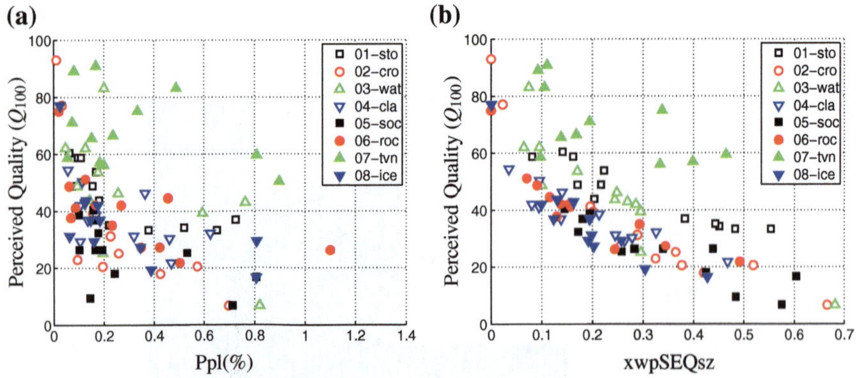

Fig. 5.24 Quality impact (on the 100-point Q_{100} scale) of the packet loss in the case of slicing using *Ppl* (percentage packet-loss) and *xwpSEQsz* as descriptors of packet loss. No content dependency is applied, i.e., $\alpha_k = 1$ for all GOPs k, in the *xwpSEQsz* computation. P.NAMS, *tr*10. **a** Ppl. **b** xwpSEQsz

temporal complexity—such as "02-*cro*" (cross-country), "05-*soc*" (soccer), "06-*roc*" (rock show), and "08-*ice*" (ice-hockey), the slicing concealment cannot work properly and losses are usually highly visible. This is reflected in Fig. 5.24b in which the above-cited temporally-complex contents are in the lower part of the plot. It has also been observed in the whole P.NAMS database set that losses are particularly visible in presence of vertical camera panning. Finally, contents "01-*sto*" and "03-*wat*" are in the middle range of the curves, therefore yielding better quality than the above-mentioned temporally complex contents, but worse quality than the TV news content. These two contents contain either spatially complex scenes (e.g complex structure in "03-*wat*" with a close shot of a fountain) or portions of picture with lots of details (as for "01-*sto*"). These two aspects may attenuate the visibility of the loss degradation.

Based on these observations, the *xwpSEQsz* parameter is improved by including a *GOP-(content)-complexity* parameter into its computation. To this aim, the following features have been studied:

- S^I_{sc}: average I-frame size per scene sc, the first I-frame of the video is ignored in the case of single-path encoding,
- S^P_{gop}: average P-frame size per GOP,
- S^B_{gop}: average reference (if any) B-frame size per GOP,
- S^b_{gop}: average non-reference B ("b") frame size per GOP,
- S^{noI}_{gop}: averaged P-, B- and b-frame sizes per GOP,
- $S^{P/I}_{gop}$: S^P_{gop}/S^I_{sc}.
- $S^{B/I}_{gop}$: S^B_{gop}/S^I_{sc}.
- $S^{b/I}_{gop}$: S^b_{gop}/S^I_{sc}.
- $S^{noI/I}_{gop}$: S^{noI}_{gop}/S^I_{sc}.

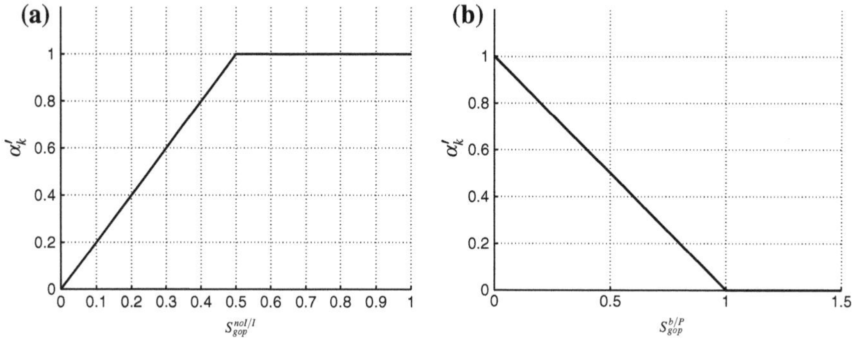

Fig. 5.25 Illustration of the computation of α'_k and α''_k from Eqs. (5.14)–(5.17). **a** α'_k. **b** α''_k

- $S_{gop}^{b/P}$: S_{gop}^{b}/S_{gop}^{P}.

Here, the index "gop" refers to the entire GOP, "sc" to the scene. All features apart from S_{sc}^{I} are computed per GOP. The objective is to include the most relevant feature(s) into the α_k coefficients of Eq. (5.6). The loss visibility is therefore adjusted on a GOP basis.

The same type of analysis as in Sect. 5.3.1 has been conducted on the frame-size related parameters, and it turned out that the most relevant features for capturing the content-dependency in the case of slicing are $S_{gop}^{noI/I}$ and $S_{gop}^{b/P}$.

These features are modulating the α_k values of Eq. (5.6) as follows:

$$\alpha_k = \alpha'_k \cdot w_1 + \alpha''_k \cdot w_2 \tag{5.13}$$

with $w_1 = w_2 = 0.5$.

For each GOP k, and as shown in Fig. 5.25a,

$$\alpha'_k = S_{gop}^{noI/I}, \tag{5.14}$$

if $\alpha'_k <= 0.5$,

$$\alpha'_k = 2 \cdot \alpha_k, \tag{5.15}$$

else

$$\alpha'_k = 1. \tag{5.16}$$

For each GOP k, and as shown in Fig. 5.25b:

$$\alpha''_k = max(0, 1 - S_{gop}^{b/P}) \tag{5.17}$$

Figure 5.26a shows the perceived quality as a function of *xwpSEQsz* for the *tr*10 database when the α_k coefficients are considered so as to be content-dependent, using Eq. (5.13). This figure is to be compared to Fig. 5.26b which corresponds to the same

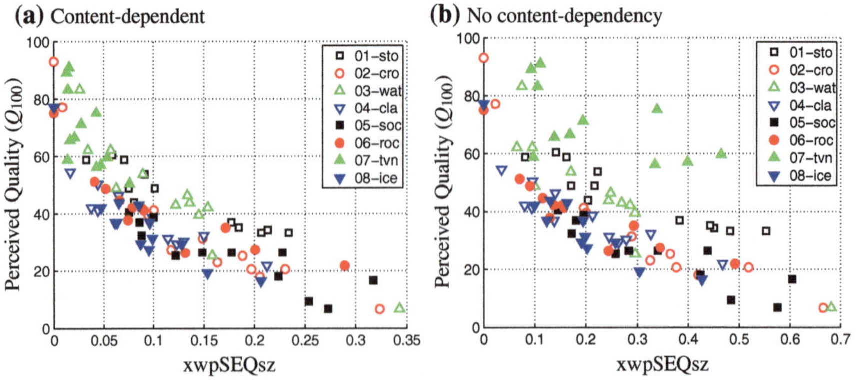

Fig. 5.26 Perceived quality as a function of *xwpSEQsz*, with and without content-dependency. P.NAMS, *tr*10. **a** Content-dependent. **b** No content-dependency ($\alpha_k = 1$)

xwpSEQsz computation but with no content-dependency ($\alpha_k = 1$ for all GOPs k). With the use of the content-dependent parameter, the curve becomes much smoother. The parameter captures particularly well the low complexity of the "07-*tvn*" content.

5.3.2.3 Degradation Pooling

Three additional steps are completed for further improving the computation of *xwpSEQsz*, the magnitude of the loss degradation in the case of slicing.

It is first expected that the more corruption there is in a GOP k, the lower the impact of additional loss contribution in GOP k. For the case of more than one slice per frame, this observation is captured by modifying Eq. (5.9) as follows:

$$xle_i = \frac{nlp_i}{np_i} + \frac{nle}{2 \cdot ns} + 0.1 \cdot ft_{log}(1 - XLE_{i-1}, -2), \tag{5.18}$$

where the function ft_{log} is defined in (Pettersson and Rossholm 2010) as shown in Eq. (5.19):

$$\begin{aligned} ft_{log} &: \mathbb{R} \times \mathbb{R} \to \mathbb{R} \\ (x, d) &\mapsto y := \tfrac{1}{d} \cdot log_{10}(1 + x \cdot (d^{10} - 1)) \end{aligned} \tag{5.19}$$

Equation (5.19) is displayed in Fig. 5.27. For the current loss event i, XLE_{i-1} represents the cumulative spatial extent of the loss events prior to i in the GOP. As a consequence, $1 - XLE_{i-1}$ represents the spatial extent of the non-degraded area before loss event i. The higher the degraded area, the smaller the non-degraded area, and the smaller the impact of the additional loss event i.

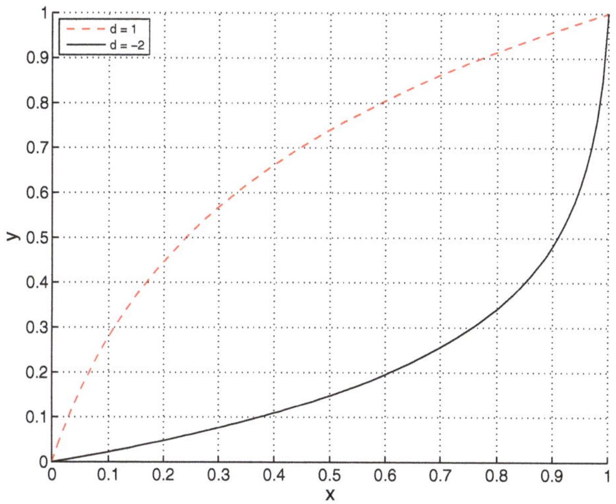

Fig. 5.27 Illustration of Eq. (5.19), see also (Pettersson and Rossholm 2010), for $d = -2$, as used in Eq. (5.18), and for $d = 1$, as used in Eq. (5.20)

The spatial extent of the loss is further transformed for reflecting the highest quality impact due to additional loss events when the current spatial extent of the loss is small. Moreover, care is taken for having a cumulative spatial extent $XLE_i \leq 1$. xle_i becomes:

$$xle_i = min(1, ft_{log}(xle_i + XLE_{i-1}, 1)) - XLE_{i-1} \tag{5.20}$$

where xle_i is computed using Eq. (5.11) in the case of one slice per frame, and using Eq. (5.18) for more than one slice per frame. The function ft_{log} has previously been defined in Eq. (5.19) and is plotted in red in Fig. 5.27 for $d = 1$.

For reflecting the observation made by Gros et al in (2001) that subjects have a longer reaction time when quality improves compared to quality decrease, the following transformation of the xl_k values is applied:

If $xl_k > xk_{k-1}$, $\tau = 0.02$, otherwise $\tau = 0.5$,

and

$$xl_k = xl_k - \frac{\tau}{T_k} \cdot (xl_k - xl_{k-1}) \cdot (1 - \exp(\frac{-T_k}{\tau})) \tag{5.21}$$

with this transformation, there is a slow improvement of the perceived quality when the degradation decreases between two GOPs (i.e., $xl_k > xl_{k+1}$), and a fast decrease of the perceived quality when the degradation increases between two GOPs (i.e., $xl_k < xl_{k+1}$).

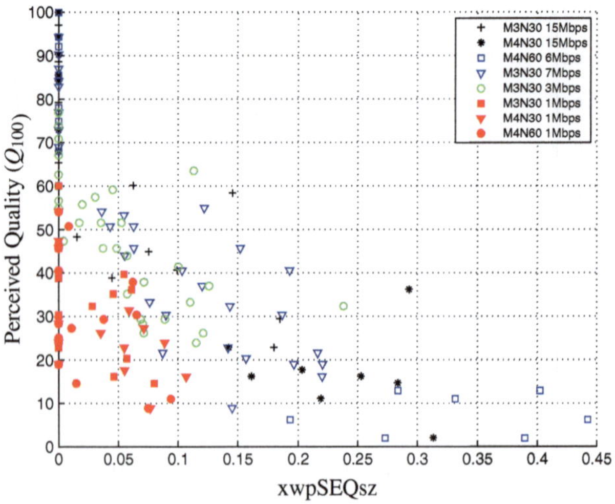

Fig. 5.28 Perceived quality as a function of *xwpSEQsz*, with content-dependency and for low to high bitrates and for various GOP strutures; P.NAMS, *tr*11

5.3.2.4 Interaction with Compression Artifacts

It should be noted that in *tr*10, packet losses are inserted only for medium and high bitrates, which all yield similar quality when there is no packet loss. In the case of low bitrates, compression and loss degradations are both influencing the perceived quality. This effect can be observed in Fig. 5.28 where all low bitrate cases ("1 Mbps") lead to low quality values as shown by the respective data points in the lower part of the plot. From 3 Mbps, the compression artifacts are not strong enough anymore for interacting with the packet loss degradation. The interaction between compression and loss degradation in the case of low bitrate is captured by adjusting the quality impact due to *xwpSEQsz* with a parameter representing the quality impact due to compression artifacts (*Icodn*). This step will be addressed in the modeling Sect. 5.4.2, Eq. (5.26).

Three types of GOP structures have been used in *tr*11: M3N30, M4N30, and M4N60. The GOP structure is impacting the loss duration, since a loss lasts longer with longer GOP length. This aspect is already captured by the *xwpSEQsz* parameter, which includes the loss duration in its computation. The GOP structure also impacts the bit allocation per frame. This effect is indirectly captured by the content-dependent α_k parameters, which includes frame-size related features. All these statements explain why the GOP structure does not seem to influence the perceived quality in Fig. 5.28.

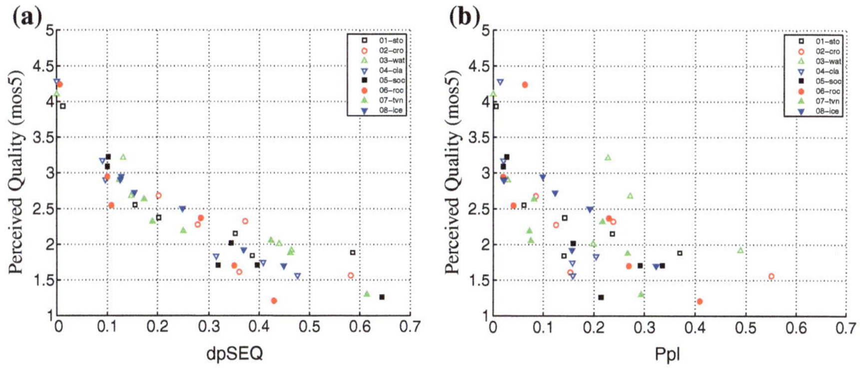

Fig. 5.29 Perceived quality as a function of *dpSEQ* and *Ppl* for *tr*10. **a** dpSEQ. **b** Ppl

5.3.3 Parametric Description of Freezing Degradation

As for *xwpSEQsz*, it is assumed that subjects perceive the freezing events occurring within a short period of time as a single degradation entity, and this entity is mapped onto quality within the judgement process. In the case of freezing, the hit frames are replaced by the previous valid frames. As a consequence, there is no "spatial extent" of the loss as such, and the perceived quality is mainly driven by the loss duration. The temporal complexity of the content is impacting the perceived quality only in the sense that the freezing event is less perceivable in static or almost static event. However, these types of contents are not dominant and the content influence stay low in the case of freezing.

As a consequence, the existence of a generic parameter describing the loss duration, *dpSEQ*, is investigated, this parameter being the overall corruption duration per measurement window normalized over the measurement window. *dpSEQ* is expressed as follows:

$$dpSEQ = \frac{fFroz}{nfr} \tag{5.22}$$

where *fFroz* is the overall number of frozen frames in the measurement window, ignoring loss starting on non-reference *b* frames, and *nfr* is the overall number of frames in the measurement window.

Figure 5.29a shows that *dpSEQ* is well describing the quality impact due to packet loss in the case of freezing. The perceived quality as a function of the "Percentage packet-loss" (*Ppl*) is shown in Fig. 5.29b for comparison.

5.4 Video Quality Model for Quality Monitoring

5.4.1 State of the Art

In this section, a parametric packet- (frame-) layer video quality model suitable for the quality monitoring of IPTV services is presented. As all models described in this work, and similarly to (Garcia and Raake 2010; Verscheure et al. 1999; Yamagishi and Hayashi 2008; You et al. 2009), the model does require access to the packet-headers but not to the decoded or partially-decoded bitstream. As a consequence, it has low computational complexity and is suitable for encrypted data. In addition, it does not access the original signal, confirming that the model is appropriate for real-time quality monitoring. It however improves the models presented in Garcia and Raake (2010), Verscheure et al. (1999), Yamagishi and Hayashi (2008), You et al. (2009) by using "frame-layer" information such as the type and size of the frames. Note that the propagation of the loss has also been considered in Yamada et al. (2010), and Frossard et al. go even one step further in Frossard and Verscheure (2001) by considering the spatial extent of the loss in the case of slicing. However, none of those previous studies make use of a single metric describing the combination of the spatial extent and duration of the loss for directly predicting perceived quality.

5.4.2 Frame-Based Video Model

The frame-based video quality model follows the same impairment-factor-based approach as the packet-based network planning video quality model, and is also developed on the 100-point (Q_{100}) scale. As for the network planning model, the quality scores are first transformed onto the perceptual scale: for the TVM databases, the ratings are first averaged over all subjects, then converted from the 11-point scale to the 5-point Mean Opinion Score (MOS) scale, and then transformed to the 100-point model-scale (Q_{100}) using the conversion defined in (ITU-T Recommendation G.107 2005) and reproduced in the Annex I of this book; for the P.NAMS databases, the ratings are averaged over all subjects and directly converted from the MOS-scale to the 100-point scale using the same ITU-T G.107 transform.

The frame-based model is decomposed as shown in Eq. (5.23):

$$Q_V = Qo_V - Icod_V - Itra_V \tag{5.23}$$

However, this time, the video quality model estimates $Icod_V$ and $Itra_V$ based on the frame-based parameters presented in the previous Sect. 5.3. This results in the following equations:

$$Icod_V = (a_1 + offset) \cdot exp(a_2 \cdot BitPerPixel \cdot Res) + a_3 \cdot SceneComp + a_4 \tag{5.24}$$

with *Res* = *2* for HD1080 and HD720 video, *Res* = *1* for SD video, *offset* = *10* for SD formats, *offset* = *0* for HD formats, and where *BitPerPixel* and *SceneComp* are expressed as shown in Eqs. (5.3) and (5.5).

In case freezing is used as PLC:

$$Itra_V = b_0 \cdot log(b_1 \cdot dpSEQ \cdot BitPerPixel + 1); \tag{5.25}$$

where *BitPerPixel* is the already defined averaged number of bits per pixel and *dpSEQ* is the overall corruption duration per measurement window, ignoring initial loss on non-reference *b* frames, normalized over measurement window. *dpSEQ* has already been described in Sect. 5.3.3.

In case slicing is used as Packet-Loss-Concealment (PLC):

$$Itra_V = c_0 \cdot log(c_1 \cdot \frac{xwpSEQsz}{Icodn} + 1) \tag{5.26}$$

with $Icodn = 1$ if $Icod < 20$ and $Icodn = 0.1125 \cdot Icod - 1.25$ otherwise.

As discussed in Sect. 5.3.2, the $\frac{xwpSEQsz}{Icodn}$ is reflecting the interaction between the quality impact due to transmission degradations—captured by the *xwpSEQsz* term—and compression artifacts—captured by the *Icodn* term—for low bitrates, i.e., for $Icod \geqslant 20$.

All coefficients a_1 to a_4, b_0, b_1, c_0 and c_1 are obtained by applying multiple non-linear regression analysis using the least-square method and the transformed subjective test results as target values.

Coefficient values are provided in Table 5.7, Sect. 5.4.3.

5.4.3 Model Evaluation

As for the audio quality model and the planning video quality model, a cross-validation procedure was adopted for evaluating the frame-based video quality model and preventing for overtraining. Since this model was submitted as video quality module to the P.NAMS competition (higher resolution application area), it is in addition evaluated against eight unknown databases produced by the P.NAMS participants and corresponding to databases *vl*13 to *vl*20 in Table 3.3.

5.4.3.1 Cross-Validation

The same cross-validation (CV) procedure as for the network planning model was applied on the five P.NAMS training video databases (*tr*08, *tr*09, *tr*10, *tr*11, and *tr*15) for the quality monitoring model. At each permutation, the model is evaluated against subjective test results in terms of Pearson Correlation Coefficient (*R*), Root-Mean-Squared-Error (*RMSE*, on the 5-point MOS scale), and epsilon-sensitive-*RMSE** (also on the 5-point MOS scale).

Table 5.5 Cross-validation coefficients for the frame-based video quality model

	AVG	STD	MIN	MAX
a_1	48.89	3.70	40.93	55.89
a_2	−10.95	1.45	−14.97	−8.81
a_3	7.48	0.84	6.26	9.57
a_4	7.82	0.82	6.53	9.68
b_0	11.55	0.66	9.96	12.65
b_1	1,039.25	272.57	622.49	1,816.93
c_0	19.40	1.75	16.85	23.62
c_1	102.81	28.08	57.40	163.13

AVG average over all CV runs; *STD* Standard deviation; *MIN* Minimum; *MAX* Maximum a_x, b_x, c_x: regression coefficients obtained for $Icod_V$, $Itra_V$ in the case of freezing, and $Itra_V$ in the case of slicing (all slicing configurations) respectively

Table 5.6 Cross-validation performance results

	R	RMSE	RMSE*
AVG	0.88	0.52	0.33
STD	0.01	0.03	0.03
MIN	0.85	0.48	0.29
MAX	0.91	0.60	0.41

AVG average over all CV runs; *STD* Standard deviation; *MIN* Minimum; *MAX* Maximum

As for the planning model, a model is considered as overtrained when at least one of its coefficients take negative and positive values for different iterations of the CV. In that case, the CV is rerun on the model without the parameter corresponding to the most unstable coefficient and performance indicators are re-computed. A "large" variation of the Pearson correlation coefficients and RMSE over the different iterations is also an indication of instability of the model. How "small" the variation should be is let to the decision of the modeler, based on her/his experience.

All regression coefficients were found stable and are therefore kept in the final model. The average (AVG), standard deviation (STD), minimum (MIN), and maximum (MAX) values of the regression coefficients resulting from the CV are listed in Table 5.5. The corresponding average, standard deviation, minimum, and maximum values of the performance indicators are shown in Table 5.6.

The model shows high and stable (low standard deviation of performance indicators over the different iterations) performance results with an averaged Pearson Correlation Coefficient of $R = 0.88$, and an averaged RMSE of $RMSE = 0.52$.

5.4.3.2 Validation on Unknown Data

The model is now trained on all P.NAMS training data (*tr*08, *tr*09, *tr*10, *tr*11, and *tr*15, i.e., 1149 video files) and evaluated on the eight validation P.NAMS database set (*vl*13 to *vl*20, 1,920 video files). The model outputs an estimate of the video quality

Table 5.7 Regression coefficients of the Quality monitoring model; coefficients are valid for all SD and HD formats

a_1	a_2	a_3	a_4
48.74	−10.85	7.45	7.82
b_0	b_1		
11.70	961.35		
c_0	c_1		
19.47	96.34		

a_x, b_x, c_x: regression coefficients obtained for $Icod_V$, $Itra_V$ in the case of freezing, and $Itra_V$ in the case of slicing (all slicing configurations) respectively

on the 100-point model-scale. The model outputs are first transformed to the 5-point scale as shown in Sect. 3.5.3. Then, a first order-polynomial mapping is applied per database on the transformed outputs of the model, and using the subjective test ratings (averaged over all subjects) as target value.

The training results in the coefficients shown in Table 5.7.

Since the resolution (i.e., number of pixels per frame) and frame rates are included in the computation of the model input parameters, these coefficients are valid for all SD and HD video formats. Similarly, the number of slices per frame is included in the calculation of the slicing degradation parameter *xwpSEQsz*, making the model independent of the slicing configuration. Finally, since parameters are computed at frame-level, the model is valid for various GOP structures as well. The number of coefficients is therefore reduced to its minimum while keeping the model valid for video formats, slicing configurations, and GOP structures not used for training the model.

Also essential, and as can be observed in Fig. 5.30 and Table 5.8, the model shows high performance results, with a Pearson correlation coefficient of $R = 0.91$, a Root-Mean-Square-Error of $RMSE = 0.44$, and an epsilon-modified RMSE value of $RMSE^* = 0.26$. This model outperforms the video quality model developed for network planning. There is still a saturation effect in the high quality range. The model is indeed capturing the content dependency only in the case of slicing, and for low and medium bitrates in the case of error-free conditions (i.e., no loss). For high bitrates and error-free conditions, the content dependency is linked to the quality of the source (original) signal, i.e., before transmitting the signal over IP. This quality variation has been kept as low as possible by selecting high quality source contents in the test databases. However, a small variation still exists, which cannot be captured by a packet- or frame-based model.

Table 5.8 further details the result per degradation type ("NoLoss", "Slicing", and "Freezing"). Best performance results are obtained for the freezing case, with an RMSE of 0.37. It is the easiest case to model, since content dependency remains low and degradation is dominated by a single factor, the loss duration. In the case of slicing or in the absence of packet-loss, the content dependency is more pronounced. It is captured by frame-type and frame-size related parameters, which allows the

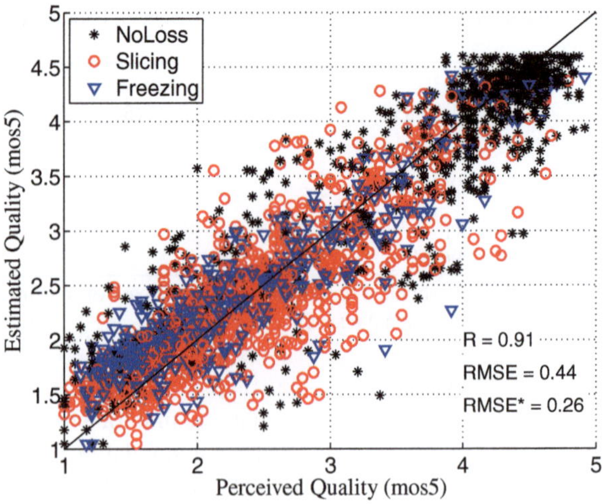

Fig. 5.30 Performance of the quality monitoring model for unknown data

Table 5.8 Performance results on unknown databases of the frame-based video quality model for all degradation types and per degradation type

Degradation	R	RMSE	RMSE*
All	0.91	0.44	0.26
NoLoss	0.87	0.46	0.28
Slicing	0.79	0.44	0.25
Freezing	0.89	0.37	0.18

Degradation types are "NoLoss" Performance results on error-free conditions (i.e., without packet-loss) "Slicing" Performance results on slicing conditions; "Freezing" Performance results on freezing conditions

model to get high performance results in the "NoLoss" and "Slicing" cases as well, with an RMSE of 0.46 and 0.44 respectively. However, since the level of information extracted from the encrypted bitstream is limited, the model cannot capture completely the content dependency. As a consequence, it does not perform as well in the "NoLoss'" and "Slicing" cases as in the "freezing" case. In the case of slicing, the model performance is also limited by the impossibility of accurately knowing at which location (horizontal and/or vertical positions) the loss occurred in the frame. This info can roughly be estimated from the rank of the lost packet in the set of packets used for encoding the frame. However, modeling attempts made with this additional information did not improve model performances. The spatial evolution (expend or reduction) of the original degradation in the dependent frames is also complex to derive from the frame-based information. The spatial evolution mainly depends on the temporal complexity of the content, and it has been partially—but probably not sufficiently—captured by the frame-size related parameters during the computation of the slicing parameter *xwpSEQsz*.

From all these observations, it can be assumed that the model has reached the highest performance results a packet- (frame-) based model can achieve. The proposed frame-based video quality model was therefore submitted as video quality module to the ITU P.NAMS competition for Higher Resolution applications. It won the competition and is now standardized under the ITU-T P.1201.2 Recommendation (ITU-T Recommendation P.1201.2 2012).

5.5 Summary

Based on the results of numerous subjective tests, the quality impact due to video compression and transmission errors has been analysed, mainly for H.264. The H.264 codec has been compared to the MPEG-2 codec. It shows higher efficiency for HD at all bitrates, and for SD at low and medium bitrates.

Based on the analysis of quality ratings for thirteen video contents, the influence on the spatio-temporal complexity of the content on the perceived quality has been studied. For low and medium bitrates, and as expected, high spatio-temporal complexity contents yield lower quality than contents with low spatio-temporal complexity. It has be observed as well that, for both SD and HD video formats, and from a key-frame-rate of 2 key-frames per second, the quality linearly decreases with an increase of the key-frame-rate (and therefore a decrease of the GOP length). This decrease is more pronounced for SD, and for spatio-temporally complex contents.

Elements influencing the perceived quality in the case of video packet loss have been extensively analysed as well. These elements include the amount of packet-loss, the packet-loss-pattern, the GOP-structure, the frame type of the frame hit by loss, the amount of packets per frame, the packet-loss-concealment, the number of slices per frame, and, finally, the spatio-temporal complexity of the content, which impacts the amount of packets per frame. Both random and burty losses have been studied when slicing or freezing is applied, and for different slicing configurations, mainly three-slices per frame and one slice per MacroBlock row. In the case of random loss, and when one slice per MB-row is used as slicing configuration, slicing yields much higher quality than freezing. It is therefore more robust to packet-loss: It achieves the same perceived quality as for freezing, but with a percentage of packet-loss eight times higher. As expected, the quality difference between slicing and freezing decreases with the decrease of the number of slices per frame. Moreover, the quality impact due to the slicing configuration decreases with the increase of burst length. Also noteworthy, the perceived quality in case of packet-loss, and for slicing and freezing, is higher with low bitrates (compared to high bitrates). This effect is smaller in the case of slicing.

All the above elements are directly or indirectly impacting the perceived quality. As a consequence, they have to be included in the model. In the case of network planning, case for which the bitstream is not available, the video quality model takes as input the video codec, the video resolution, the bitrate, the frame-rate, the key-frame-rate, the percentage of packet-loss, the average burst length, the packet-loss-concealment type, and the number of slices per frame. In the case of quality monitoring, case for which video frame-related information can be obtained, the model takes as inputs the codec, the resolution, the frame rate, the bitrate, a parametric description of the content complexity, the packet-loss-concealment type, a parametric description of the freezing degradation, mainly capturing the overall loss duration, and a parametric description of the slicing degradation capturing the overall spatial extent and loss duration of the degradation. In the case of slicing, the influence of the content in terms of loss visibility is also considered and derived from frame-type and frame-sizes information.

Both models output an estimate of the perceived video quality on the 100-point model-scale, which can be transformed onto the 5-point MOS scale. The models have been first evaluated using a cross-validation procedure, then using unknown subjective test data.

The network planning packet-based model obtains on unknown data a Pearson Correlation coefficient of $R = 0.84$ with subjective tests ratings and a $RMSE = 0.56$. The quality monitoring frame-based model shows higher performance results with $R = 0.91$ and $RMSE = 0.44$.

Note that for the frame-based model, a single set of coefficients is valid for all video formats, frame-rates, slicing configurations, and GOP structures. This model corresponds to the video quality module of the "T-Labs" model sent to the ITU P.NAMS competition for the Higher Resolution applications area. It won the competition and became part of the ITU-T P.1201 standard (ITU-T Recommendation P.1201.2 2012) (see also Garcia et al. 2013).

Chapter 6
Audiovisual Quality Model

This chapter presents the parametric audiovisual quality model developed in this work. The model estimates the perceived audiovisual quality of Standard Definition (SD) and High Definition (HD1080, HD720) audiovisual streams. The same model is used for all SD and HD video formats. It is developed for application to network planning and quality monitoring, and is suitable for encrypted audiovisual streams. It covers mainly audio and video compression artifacts and impairments due to packet loss. Addressed audiovisual contents are typical TV programs such as sports and movies. Subjective tests conducted for developing the audiovisual quality model have already been presented in Chap. 3 and are referred to as the TVM, P.NAMS training and P.NAMS validation database sets. They will briefly be summarized in the first section (Sect. 6.1) of this chapter before analysing the results. It will be shown that there is a mutual influence of the perceived audio and video qualities, and a predominance of the video quality for the overall audiovisual quality, especially in the HD case. The balance between audio quality and video quality however depends on the content and video format (SD versus HD).

The proposed model—referred to as "Impairment-Factor-based" model, or "IF-based" model—is based on impairment factors which quantify the quality impact due to the different degradations, reflecting the observation that certain types of degradations affect quality differently in an audio- or video-only context than in an audiovisual context. The impairment factors are taken from the parametric packet-based audio and video quality models presented in Chaps. 4 and 5. In the case of network planning application, the packet-based video quality model developed for network planning is used; in the case of quality monitoring, the frame-based video quality model is used instead.

For comparison, a more classical audiovisual quality model—referred to as "Quality-based" ("Q-based") model—has been developed. It is based on the audio and video qualities and their interaction. Due to a more suitable test design, the two modeling approaches will be compared using the TVM set (Sect. 6.2). The degradation-type- and content-dependencies will also be discussed. Further, the different model variants are compared with modeling approaches described in the literature.

M.-N. Garcia, *Parametric Packet-Based Audiovisual Quality Model for IPTV Services*, T-Labs Series in Telecommunication Services, DOI: 10.1007/978-3-319-04855-0_6,

As for the audio and video quality models, the audiovisual quality model was ultimately developed in the context of the P.NAMS competition. The two models ("IF-based" and "Q-based") are therefore eventually trained on the P.NAMS training database set (Sect. 6.3). A cross-validation procedure is then run on the P.NAMS training set for validating the models and preventing for overtraining. Finally, the models are validated on the P.NAMS validation set.

The IF-based model predictions show high performance results on unknown data, with a correlation of 0.92 with subjective scores, and a root-mean-square error of 0.42 on the 5-point scale used in the subjective tests.

6.1 Subjective Quality Assessment

6.1.1 Subjective Test Overview

The TVM and P.NAMS audiovisual databases have already been presented in Sect. 3.3. Mains aspects on the experimental design are reminded in this section for reading convenience purpose.

Audio, video and audiovisual subjective tests have been conducted using audio-only, video-only, and audiovisual sequences respectively. Video-only and audiovisual tests were conducted separately for each video format. In the TVM databases, the audio and video stimuli used in the audio-only and video-only tests correspond to the audio and video of the audiovisual stimuli used in the audiovisual tests. As a consequence, for each audiovisual stimulus, the perceived audio-only, video-only, and audiovisual qualities are known. They will be referred to as the "ground truth". This is not the case for the P.NAMS data, in which the audio-only and video-only qualities of the audiovisual stimuli can only be estimated using audio and video quality prediction models.

As presented in Sect. 3.3, the audiovisual quality tests were conducted for collecting opinion ratings, using an ACR test method with the continuous 11-point scale (ITU-T Recommendation P.910 1999) in the TVM database set, and the 5-point categorical scale in the P.NAMS database sets. Respectively four, three, and four tests were conducted in the TVM, P.NAMS training and validation database sets. As described in the subsection "Test Design" of Sects. 3.3.2 and 3.3.3, the same anchors (Tables 3.11 and 3.23) were used in all subjective tests for a given database set (TVM or P.NAMS). These anchors cover the whole quality range and degradation types used in the tests. They are used for mapping the results of the different tests, as described in Sect. 3.4.2.

Four audiovisual tests were conducted for the TVM set, $ph2^{SD}_{AV}$, $ph2^{HD}_{AV}$, $ph3^{SD}_{AV}$, and $ph3^{HD}_{AV}$. The same conditions, listed in Table 3.10 , were applied in $ph2$ and $ph3$ on the same five contents, listed in Table 3.4. However, due to processing issues, results for contents A and D were not used in $ph2^{[H,S]D}_{AV}$, and freezing packet loss concealment was present only in the anchor conditions, making the $ph2^{[H,S]D}_{AV}$

databases still balanced in terms of quality range and perceptual dimensions, but the ratings for freezing conditions and contents A and D could not be used for evaluating the model. Further note that loss processing between the *ph*2 and *ph*3 databases was done independently, yielding different loss instances in the decoded audio and video.

It should be reminded that the audio and video streams of the five audiovisual TVM source contents were processed off-line separately, and the test conditions listed in Table 3.10 are representative of typical IPTV degradations. As is typical of IPTV services, an MPEG2-TS/RTP/UDP/IP packetization scheme was used. Here, seven MPEG-2 transport stream (TS) packets are contained in one RTP packet, and each contains either audio or video. For the TVM tests, multiplexing was done for the already decoded files, instead of using ecologically valid multiplexing at TS-level. Note that this choice was made to ensure that the resulting model will be valid in a variety of situations with different levels of audio- and video-degradations. This is reflected especially in the combinations of loss rates, where different settings have been used for audio and video.

The audio stream of the audiovisual stimuli were taken from the audio tests *ph*2 and *ph*3, and the video streams from the video tests *test*1—*SD* and *test*1—*HD* for $ph2_{AV}^{SD}$ and $ph2_{AV}^{HD}$, and from *test*2—*SD* and *test*2—*HD* for $ph3_{AV}^{SD}$ and $ph3_{AV}^{HD}$. Note that the audio-only, video-only, and audiovisual tests span the same audio- and video-quality ranges. This avoids introducing bias in the audiovisual tests and ratings of the audio- and video-only tests can be directly used as ratings for the audio and video streams of the audiovisual stimuli.

Seven audiovisual tests were produced within P.NAMS: three databases used for training the models (*tr*12, *tr*13, *tr*14) and four databases for evaluating them (*vl*21, *vl*22, *vl*23, *vl*24). Contents used in these databases are listed in Tables 3.13 and 3.14. For these databases, a more ecologically valid approach was used for the processing, and multiplexing of audio and video was done at TS-level. In order to simulate typical IPTV degradations, the audiovisual stimuli were processed off-line for the P.NAMS databases and following the test conditions described in Table 3.17 for the P.NAMS training databases and in Table 3.20 for the P.NAMS validation databases. As for TVM, subjective tests were conducted separately for each video format.

6.1.2 Subjective Test Results

As previously mentioned, the ground truth for the perceived audio-only, video-only, and audiovisual qualities is known in the TVM audiovisual tests only, allowing a more accurate analysis of the relation between audio, video, and audiovisual qualities. As a consequence, the subjective test analysis will be conducted on the TVM data.

For each of the TVM subjective tests, the scores were averaged over all subjects, yielding Mean Opinion Scores (MOS). These MOSs were linearly transformed to the 5-point ACR MOS scale by aligning the numbers of the scales, and further transformed to the 100-point model-scale using the conversion defined in ITU-T

Table 6.1 Correlation of the audio and video quality, and their interaction with the overall quality (SD and HD)

	Q_{AV}	All	A	B	C	D	E
HD	Q_A	0.47	0.49	0.46	0.45	0.45	*0.57*
	Q_V	0.80	0.87	0.80	0.86	0.84	*0.69*
	$Q_A \cdot Q_V$	0.92	0.94	0.94	0.94	0.95	*0.92*
SD	Q_A	0.51	0.57	0.52	0.48	0.48	*0.61*
	Q_V	0.75	0.73	0.77	*0.81*	0.79	*0.67*
	$Q_A \cdot Q_V$	0.94	0.94	0.95	0.92	0.96	*0.96*

Recommendation G.107 (ITU-T Recommendation G.107 2005) and reproduced in Annex I.

Note that in the following, and unlike (Garcia and Raake 2009), the ratings are not averaged across all contents but per content. This choice is motivated by two reasons: (a) the audiovisual quality model is to be applied on audiovisual sequences with various contents, and a predicted quality value per sequence is required; the quality variation due to content should therefore be captured. (b) The audiovisual quality model developed for all contents, i.e., with one set of coefficients valid for all contents, will be compared to an audiovisual quality model with different sets of coefficients for each content.

In order to have a first impression of the quality impact due to audio and video on the overall audiovisual quality, a correlation analysis was conducted, correlating the audio-quality Q_A, the video-quality Q_V, and their interaction $Q_A \cdot Q_V$ with the audiovisual quality Q_{AV} (see Table 6.1, column "all"). It can be observed that for both SD and HD, the interaction term is predominant (SD: correlation $= 0.94$; HD: correlation $= 0.92$). The video quality seems to have more impact on the overall audiovisual quality than the audio quality, especially for HD (SD: video correlation $= 0.75$, audio correlation $= 0.51$; HD: video correlation $= 0.80$, audio correlation $= 0.47$). This finding is expected, and it shows that the impact of video quality increases with the video format.

Moreover, the individual impact of the audio and of the video quality on the overall audiovisual quality depends on the quality of the other modality (video and audio). This is reflected by the slopes of the edges in Fig. 6.1: the audio quality Q_A has a decreasing influence on the overall HD audiovisual quality Q_{AV} for decreasing HD video quality. In turn, the HD video quality Q_V has a less strongly declining influence on the overall HD audiovisual quality with decreasing audio quality. Similar observations have been made for SD. Note that, for the sake of clarity, Fig. 6.1 shows the ratings averaged over all subjects and over all contents, that is the per-condition ratings instead of the per-sequence ratings.

Using the results for all contents might hide that for some contents, the above statements are not valid anymore. As a consequence, the same correlations as above were computed, but using ratings per content (see Table 6.1, columns "A" to "E").

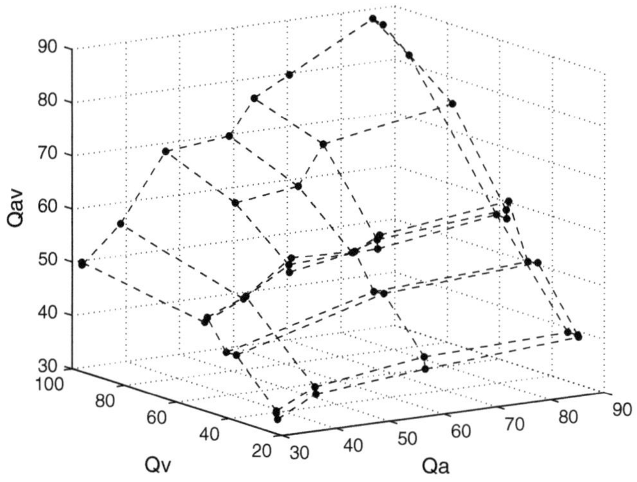

Fig. 6.1 Perceived audiovisual quality (Q_{AV}) as a function of the perceived audio (Q_A) and video quality (Q_V). HD resolution

For the contents "A" to "D", the same observation as for "all" contents can be made. For content "E" (music video), the quality impact due to audio seems to be higher than for the other contents, and closer to the quality impact due to video (SD: correlation $(Q_{AV}, Q_A) = 0.61$; HD: correlation $(Q_{AV}, Q_A) = 0.57$). This observation especially applies to SD, confirming the impact of the video format.

It should be noted that the influence of the content on the perceived quality plays a role at different levels. For instance, in the video-only case, it is well-known that the quality impact due to the bitrate is highly content-dependent (Péchard et al. 2007; Liu et al. 2006; Ries et al. 2007; Khan et al. 2009; Garcia et al. 2010), especially at low bitrates. This result can be captured by developing video quality models that are explicitly taking video content characteristics into account, as is proposed in the frame-based video quality model, Sect. 5.4.2. In this chapter, focus is given on the influence of the audiovisual content on the balance between audio and video quality, and on how this variation can be captured in the audiovisual quality modeling.

One more aspect to be considered is how the degradation type influences the quality impact due to audio and video on the overall audiovisual quality. In the present case, the employed degradation types were audio and video compression, audio frame loss, and video packet loss. It should be studied if, for instance, for a given level of audio and video qualities, different audiovisual quality values for audio compression and for audio packet loss are obtained, even though both have resulted in the same audio-only quality in the audio test.

Both the quality impact due to the content type and to the degradation type are discussed in the following section.

6.2 Modeling Approach

6.2.1 State of the Art

Several studies on audiovisual perception have been conducted starting in the 80ties [summarized in Kohlrauch (Kohlrausch and van de Par 2005)]. However, the first audiovisual quality models to be found in the literature appeared as late as in the 90ties. At this time, they addressed either analog degradations such as audio and video noise—this is the case for Bellcore's (ITU-T SG12 1993; ITU-R SG12 1994) and Beerends' models (Beerends and Caluwe 1999)—or compression artifacts such as blockiness—this is the case for France Telecom's (Chateau 1998), NTIA-ITS' (ITU-T SG12 D.038 1998; Jones and Atkinson 1998), and Hands' (Hands 2004) models. For an overview of audiovisual quality models covering analog and compression degradations, see (You et al. 2010). The interest in modeling audiovisual quality is currently rising again, reflected for instance by standardization activities such as the ITU-T P.1201 Rec. or the Audiovisual High Definition Quality (AVHD) project of the Video Quality Expert's Group (VQEG), which intends to evaluate audiovisual quality models for multimedia applications.[1] Audiovisual quality models for mobile applications have been developed in 2005 and 2006 by Ries et al. (2005) and Winkler (Winkler and Faller 2006), but the reported model versions do not cover the effect of transmission errors. This latter point is problematic since in the case of the time-varying degradation due to transmission errors, the impact of audio and video quality on the overall audiovisual quality as well as their interaction might differ from the case of compression artifacts. In Belmudez et al. (2009) address the impact of transmission errors in addition to compression and frame rate artifacts, but for interactive scenarios and small video resolutions, which is not suitable for our application. Pinson et al. provide in (Nov. 2011) an interesting overview of existing audiovisual quality models. They show that a simple model based on the product of audio and video quality terms is valid for a wide range of scenarios and applications. When a small amount of data is available for a wide range of application, it is indeed a safe choice to use a model as simple and with as few coefficients as possible for avoiding overtraining. At the time T-Labs needed an audiovisual quality model for High Bitrate applications such as IPTV and VoD, no models were available in the literature for such applications. As a consequence, a new audiovisual quality model which covers all IPTV-typical degradations—mainly audio and video compression artifacts and packet loss—and which is applicable to both SD and HD formats had to be developed. Based on the quality perception tests conducted during model development, the influence of the degradation type and of the audiovisual content on the quality impact due to audio and video has been analysed.

[1] Unfortunately, to the knowledge of the authors, no citable document describing the VQEG AVHD has been published at the time of writing this book.

For modeling audiovisual quality, the impairment-factor-based approach described in Sect. 3.5.1 is followed. The audiovisual quality is therefore computed from audio and video impairment factors instead of audio and video qualities, as is done in most previous studies. NTT followed a similar approach in (Yamagishi and Hayashi 2005), but their model has been developed for interactive multimedia services such as video-telephony, yielding psychological factors not applicable in the case of IPTV, such as "feeling of activity".

As mentioned in the introduction, the Impairment-Factor-based approach is compared to the more classical quality-based approach, which is based on the audio and video qualities and their interaction.

6.2.2 Modeling Approach

Two approaches for modeling the audiovisual quality will be presented in this section. The first approach was used in previous studies and consists in modeling the audiovisual quality from the audio-only Q_A and the video-only Q_V quality terms, as shown in Eq. (6.1):

$$Q_{AV} = \alpha + \beta \cdot Q_A + \gamma \cdot Q_V + \zeta \cdot Q_A \cdot Q_V \tag{6.1}$$

The second approach consists in modeling the audiovisual quality from the audio and video quality factor terms $Icod_A$, $Icod_V$, $Itra_A$, and $Itra_V$, as shown in Eq. (6.2):

$$\begin{aligned} Q_{AV} = {} & Qo_{AV} \\ & - c_{ac} \cdot Icod_A - c_{at} \cdot Itra_A \\ & - c_{vc} \cdot Icod_V - c_{vt} \cdot Itra_V \\ & - c_{avcc} \cdot Icod_A \cdot Icod_V - c_{avtt} \cdot Itra_A \cdot Itra_V \\ & - c_{avtc} \cdot Itra_A \cdot Icod_V - c_{avct} \cdot Icod_A \cdot Itra_V \end{aligned} \tag{6.2}$$

where Qo_{AV} is the maximum audiovisual quality on the modeling scale (e.g., $Qo_{AV} = 100$), $Icod_X$ is the quality impact due to video ($X \equiv V$) or audio ($X \equiv A$) compression, and $Itra_X$ is the quality impact due to video- or audio-transmission errors.

It should be reminded that the audio or video quality is related to the audio or video impairment factors using the following relation:

$$Q_X = Qo_X - Icod_X - Itra_X, \tag{6.3}$$

wherein Qo_X is the maximum quality rating on the modeling scale (e.g $Qo_X = 100$). As a consequence, $Icod_X$ is derived from the subjective tests for transmission error-free conditions as follows: $Icod_X = Qo_X - Q_X$. $Itra_X$ is obtained using all conditions by computing $Itra_X = Qo_X - Icod_X - Q_X$.

Both models will be trained using the TVM subjective scores (*ph*2 and *ph*3 for audio, *test*2—*SD* and *test*2—*HD* for video) for deriving the audio and video quality terms Q_X and impairment factor terms $Itra_X$. Indeed, using audio and video quality terms predicted by instrumental models would introduce biases—the models' prediction errors—which may modify the coefficients of the audiovisual quality model obtained during the modeling, and therefore bias the analysis of the modeling approaches. Using the TVM data set, for which the subjective audio, video, and audiovisual qualities (the "ground truthes") are known for each audiovisual stimuli, is therefore the most valid available way of comparing and evaluating the two modeling approaches.

Once validated, the two modeling approaches will be re-used in Sect. 6.3.1 for developing four variants of the audiovisual quality model. However, this time, and for better reflecting the real instrumental assessment scenario, the audio and video quality terms and impairment factors will be derived from the audio and video models presented in the previous chapters, and not from subjective scores.

6.2.3 Quality-Based Modeling

6.2.3.1 Model

Similarly to other studies (ITU-T SG12 1993; ITU-R SG12 1994; ITU-T SG12 D.038 1998; Jones and Atkinson 1998; Beerends and Caluwe 1999; Chateau 1998; Hands 2004; Ries et al. 2005; Winkler and Faller 2006; Yamagishi and Hayashi 2005), the audiovisual quality Q_{AV} is now modeled based on the audio quality Q_A, the video quality Q_V, and the interaction between Q_A and Q_V, as shown in Eq. (6.4):

$$Q_{AV} = \alpha + \beta \cdot Q_A + \gamma \cdot Q_V + \zeta \cdot Q_A \cdot Q_V \tag{6.4}$$

This model is called "'quality-based" model, or "Q-based" model. The coefficients α, β, γ, and ζ of Eq. (6.4) vary from one research to the other, depending on the application, the resolution of the video, and on the audiovisual content. By applying the quality-based model on SD and HD ratings averaged over all subjects, and transformed onto the 100-point model-scale, the coefficients of Table 6.2, rows "all", are obtained. The content-based audiovisual quality model with different coefficients per content is obtained by applying the quality-based model to ratings averaged over all subjects for each content separately. The obtained coefficients are listed in rows "A" to "E" of Table 6.2.

The regression coefficients are compared taking into account their 95 % confidence intervals: if the confidence intervals of two regression coefficients do not overlap, the regressions coefficients are significantly different. If the confidence interval of a coefficient overlaps the value zero, the regression coefficient is considered as non-significant, i.e., not different from zero.

Table 6.2 Regression coefficients of the quality-based model for HD and SD, across all contents (rows "all") and per content (rows "A" to "E")

	α	β	γ	ζ
HD all	*28.49*	*0*	*0.13*	*0.006*
HD A	24.57	0	0.28	0.006
HD B	27.50	0	0.11	0.006
HD C	24.37	0	0.21	0.005
HD D	27.85	0	0.17	0.005
HD E	32.59	0	**0**	0.007
SD all	*30.99*	*0*	*0*	*0.006*
SD A	32.77	0	0	0.006
SD B	30.21	0	0	0.006
SD C	25.83	0	**0.15**	0.005
SD D	32.06	0	0	0.006
SD E	30.83	0	0	0.006

In the present HD case, and similarly to (Hands 2004) for high-motion video, the dominance of the video quality over the audiovisual quality leads to $\beta = 0$. For SD, $\beta = 0$ and $\gamma = 0$, confirming that audio quality and video quality are more balanced for SD. This is in accordance with the observations made on the correlation values shown in Table 6.1, Sect. 6.1.2.

6.2.3.2 Quality Impact Due to Content

When modeling the per-content data (coefficients of rows "A" to "E"), it can be observed that the model pattern depends on the content. Indeed, for HD, γ is significantly different from zero for all contents except content E (music video). This result was expected from the observation made on the correlations between audio and video qualities (see Sect. 6.1.2, Table 6.1): the impact of audio and video quality is more balanced for content E than for the other contents. Similarly, it had been observed in Sect. 6.1.2 that the audio and video quality was more balanced for SD than for HD. This balance is less respected in the case of content C (soccer), for which the correlation between video and audiovisual qualities is higher than for the other contents. This is translated into a non-zero value of γ found in the regression analysis.

6.2.4 Impairment-Factor-Based Modeling

6.2.4.1 Model

The advantage of the quality-based model variant is that it can easily be used with audio and video quality models coming from other laboratories, provided they are based on similar types of network conditions and services, and deliver quality estimates on the same scale. The flip-side to this advantage is that the quality-based model does not allow for a fine-grained diagnosis of the cause for non-optimum quality. Indeed, using Eq. (6.4), it can only be known if a low audiovisual quality Q_{AV} is caused by a low audio quality Q_A, a low video quality Q_V, or both. For diagnostic purposes, the audio and video impairment factors $Icod_X$ and $Itra_X$ can be computed, and therefore, using Eq. (6.3), the audio- (Q_A) and video- (Q_V) quality impact due to audio and video degradations can be investigated. However, it remains unknown if these degradations have a similar impact in an audiovisual perception context. Let's insert Eq. (6.3) for both audio and video into Eq. (6.4):

$$\begin{aligned} Q_{AV} = {} & (\alpha + \beta \cdot Qo_A + \gamma \cdot Qo_V + \zeta \cdot Qo_A \cdot Qo_V) \\ & - (\beta + \zeta \cdot Qo_V) \cdot Icod_A - (\beta + \zeta \cdot Qo_V) \cdot Itra_A \\ & - (\gamma + \zeta \cdot Qo_A) \cdot Icod_V - (\gamma + \zeta \cdot Qo_A) \cdot Itra_V \\ & + \zeta \cdot Icod_A \cdot Icod_V + \zeta \cdot Itra_A \cdot Itra_V \\ & + \zeta \cdot Itra_A \cdot Icod_V + \zeta \cdot Icod_A \cdot Itra_V \end{aligned} \tag{6.5}$$

Identical coefficients in Eq. (6.5) imply a similar impact on audiovisual quality. This is, for example, the case for all interaction-terms between impairment factors $Icod_X$ and $Itra_X$, which are all multiplied by the same coefficient ζ. Thus, this model assumes that all interaction-terms between impairment factors have the same weight for audiovisual quality. Similarly, Eq. (6.5) suggests that for each modality (audio and video), the individual terms $Icod_X$ and $Itra_X$ with equal X (audio or video) have the same impact on audiovisual quality.

To verify the validity of this assumption—that will be called assumption "A" –, the audiovisual quality is directly expressed as a function of the impairment factors, using Eq. (6.6):

$$\begin{aligned} Q_{AV} = {} & Qo_{AV} \\ & -c_{ac} \cdot Icod_A - c_{at} \cdot Itra_A \\ & -c_{vc} \cdot Icod_V - c_{vt} \cdot Itra_V \\ & -c_{avcc} \cdot Icod_A \cdot Icod_V - c_{avtt} \cdot Itra_A \cdot Itra_V \\ & -c_{avtc} \cdot Itra_A \cdot Icod_V - c_{avct} \cdot Icod_A \cdot Itra_V \end{aligned} \tag{6.6}$$

The name convention for the coefficients is as follows: the subscripts a, v, c and t stand for *audio*, *video*, *coding*, and *transmission* respectively. When c and t are both

Table 6.3 Regression coefficients of the IF-based model for HD and SD, for all contents (row "all") and per content (rows "A" to "E")

	Qo_{AV}	c_{ac}	c_{vc}	c_{avcc}	c_{at}	c_{vt}	c_{avtt}	c_{avtc}	c_{avct}
HD all	*94.33*	*0.466*	*0.713*	*− 0.008*	*0.652*	*0.712*	*−0.007*	*−0.009*	*−0.007*
HD A	n.a.	n.a.	n.a.	n.a.	n.a.	n.a.	n.a.	n.a.	n.a.
HD B	94.33	0.539	0.814	−0.010	0.752	0.727	−0.009	−0.017	−0.008
HD C	94.33	0	0.786	0	0.685	0.724	−0.007	−0.012	0
HD D	94.33	0.416	0.851	−0.007	0.601	0.724	−0.007	−0.013	−0.007
HD E	94.33	0.560	0.519	−0.009	0.711	0.667	−0.008	−0.011	−0.009
SD all	*82.90*	*0.387*	*0.511*	*−0.004*	*0.539*	*0.507*	*−0.005*	*−0.006*	*−0.006*
SD A	82.90	0.333	0.411	0	0.471	0.523	−0.004	0	−0.008
SD B	82.90	0.510	0.521	−0.006	0.677	0.522	−0.004	−0.012	−0.007
SD C	82.90	0	0.657	0	0.567	0.462	−0.002	−0.010	0
SD D	82.90	0.324	0.472	−0.004	0.559	0.492	−0.005	−0.005	−0.004
SD E	82.90	0.309	0.398	0	0.613	0.484	−0.006	−0.007	0

present in the coefficient name, the first of those two letters is related to audio, the second to video. As an example, c_{avct} represents the coefficient of the interaction between the *audio coding* impairment $Icod_A$ and the *video transmission* impairment $Itra_V$.

Note that the interactions between $Icod_A$ and $Itra_A$, and between $Icod_V$ and $Itra_V$ are implicitly taken into account by including them in $Itra_A$ and $Itra_V$ [see Eqs. (4.3), (5.4), or (5.26)]. As a consequence, Eq. (6.6) does not explicitly contain the interaction terms $Icod_A \cdot Itra_A$ and $Icod_V \cdot Itra_V$.

6.2.4.2 Quality Impact Due to Degradation Type

If the regression coefficients c_{ac} and c_{at}, or c_{vc} and c_{vt}, or c_{avcc}, c_{avtt}, c_{avtc} and c_{avct} are significantly different, assumption "A" cannot be validated, that is that the respective impairments have the same impact on overall quality. As for the quality-based model, the regression coefficients are compared taking into account their confidence intervals.

Applying multiple regression analysis using the results of the audio-only, the video-only, and the audiovisual subjective tests with Eq. (6.6), the regression coefficients shown in Table 6.3 are obtained, row "HD all" for HD, row "SD all" for SD. Due to processing issues,[2] the coefficients for the impairment-factor-based model could not be developed for content A, HD resolution.

Regression coefficients and their confidence intervals are displayed in Figs. 6.2 and 6.3 for HD and SD respectively. Significance-related information for the regression

[2] Only one video file, present in both the video and audiovisual tests, was corrupted. However, it was crucial for computing the $Icod_V$ value of several video files with transmission errors and having the same bitrate using the equation $Itra_V = Qo_V - Icod_V - Q_V$ as shown in Sect. 6.2.2.

Table 6.4 Significance of regression coefficients and of differences between regression coefficients, for HD and SD, from Eq. (6.6)

	Non-sign. coeff.	Sign. diff. coeff.-pairs
HD all	*None*	$\{c_{ac}, c_{at}\}$
HD A	n.a.	n.a.
HD B	None	$\{c_{ac}, c_{at}\}$
HD C	$c_{ac}, c_{avcc}, c_{avct}$	$\{c_{ac}, c_{at}\}$
		$\{c_{avcc}, c_{avtt}\}, \{c_{avcc}, c_{avtc}\}$
		$\{c_{avct}, c_{avtt}\}, \{c_{avct}, c_{avtc}\}$
HD D	None	$\{c_{ac}, c_{at}\}$
HD E	None	None
SD all	*None*	$\{c_{ac}, c_{at}\}$
SD A	c_{avcc}, c_{avtc}	$\{c_{avcc}, c_{avtt}\}, \{c_{avcc}, c_{avct}\}$
		$\{c_{avtc}, c_{avtt}\}, \{c_{avtc}, c_{avct}\}$
SD B	None	None
SD C	$c_{ac}, c_{avcc}, c_{avct}$	$\{c_{ac}, c_{at}\}, \{c_{vc}, c_{vt}\}, \{c_{avtc}, c_{avtt}\}$
		$\{c_{avcc}, c_{avtt}\}, \{c_{avcc}, c_{avtc}\}$
		$\{c_{avct}, c_{avtt}\}, \{c_{avct}, c_{avtc}\}$
SD D	None	$\{c_{ac}, c_{at}\}$
SD E	c_{avcc}, c_{avct}	$\{c_{ac}, c_{at}\}$
		$\{c_{avcc}, c_{avtt}\}, \{c_{avcc}, c_{avtc}\}$
		$\{c_{avct}, c_{avtt}\}, \{c_{avct}, c_{avtc}\}$

coefficients are shown in Table 6.4, rows "HD all" and "SD all". Coefficients not significantly different from zero and coefficients significantly different from the other coefficients are indicated in the columns "non-sign. coeff." and "sign. diff. coeff.-pairs".

It can be seen in both Figs. 6.2 and 6.3 and Table 6.4 that all regression coefficients are significantly different from zero. A remarkable behavior can be observed in the case of coefficients c_{ac} and c_{at}, which are linked to the quality impact due to audio. As becomes apparent from Figs. 6.2 and 6.3, and Table 6.4, these coefficients are statistically different both for SD and HD. This means that (a) audio quality alone shows a significant impact on audiovisual quality, when the coding and packet-loss related contributions to audio quality are separated, and (b) the impairment due to audio packet loss impacts audiovisual quality differently from that due to audio coding. Hence, when a coding-only audio impairment and a transmission-related audio impairment of equal value $Icod_A = Itra_A$ are presented to users in an audiovisual context, the packet-loss related impairment plays a larger role for audiovisual quality than the coding-related one. Both of these effects cannot be captured by the quality-based model, where the audio-only quality was not found to have a significant impact on the overall audiovisual quality. Since c_{ac} is significantly different from c_{at}, assumption "A" can be rejected, confirming that the degradation type does have an influence on how the audio component impacts audiovisual quality. This supports the idea of impairment-factor-based modeling approach.

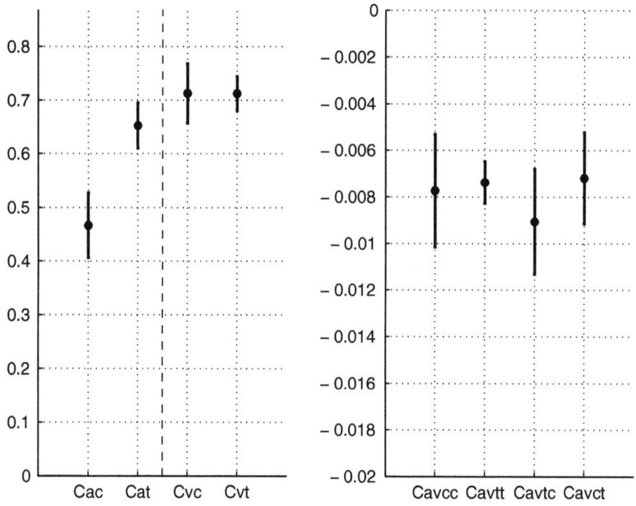

Fig. 6.2 Regression coefficients and their 95 % confidence intervals. IF-based HD model

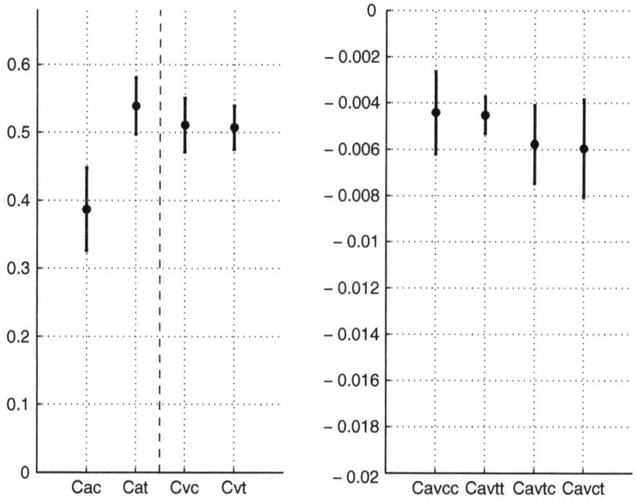

Fig. 6.3 Regression coefficients and their 95 % confidence intervals. IF-based SD model

From the analysis of their confidence intervals, it cannot be concluded that the regression coefficients of $Icod_V$ and $Itra_V$ or of all the interaction terms are different. This may mean that the impact of video on audiovisual quality is independent of the video degradation type, and that the impact of the interaction between audio and video qualities on the audiovisual quality is independent of the audio and video degradation types. This may also mean that the influence of the audio and video degradation types has been compensated by the influence of the content type during the modeling process. Indeed, if the analysis is done per content, $Icod_V$ is shown to have a higher impact on the audiovisual quality than $Itra_V$, but for another content, the opposite is observed; on average, $Icod_V$ and $Itra_V$ will have the same impact on audiovisual quality, and will therefore not have significantly different regression coefficients.

All these results provide us with interesting insights into the subjects' attention in the context of audiovisual quality assessment. Indeed, in an audiovisual test, the subjects seem to focus more on video, as in a video-only test, while the audio is only subconsciously attended to. With their main attention on the video, the subjects pay similar attention to stationary degradations such as compression artifacts as to more time-varying degradations such as transmission errors, just as in a video-only test. The users' attention is attracted more to the audio only in the case of transient audio degradations such as audio frame loss. This may explain why—across contents—the coefficients of $Icod_V$ and $Itra_V$ are not significantly different, while the coefficient of $Itra_A$ is significantly bigger than the one of $Icod_A$.

6.2.4.3 Quality Impact Due to Degradation and Content Types

For investigating the impact of the content on audiovisual quality, the regression analysis on ratings averaged per content over all subjects is re-run. The obtained regression coefficients are shown in Table 6.3, rows "A" to "E". Coefficients not significantly different from zero are shown in column "non-sign. coeff." of Table 6.4 for each resolution (referred to by "HD" and "SD"), and separately for each content (rows "A" to "E"). Moreover, it should be verified if assumption "A" still can be rejected when modeling the audiovisual quality per content. For this purpose, in column "sign. diff. coeff.-pairs" is indicated for each resolution and content, if $c_{ac} \neq c_{at}$, or $c_{vc} \neq c_{vt}$, or if one of the coefficients of the multiplicative terms of Eq. (6.6) is significantly different from any other.

It can be observed that for some of the contents some regression coefficients are non-significant (e.g., coefficient $c_{ac}, c_{avcc}, c_{avct}$ of content C for HD) but for other contents they are (e.g., contents B, D, and E for HD). This implies that different model patterns for different contents may increase the overall performance of the model. Moreover, for several contents, c_{ac} is significantly different from c_{at}, confirming that the audio-only quality does have an impact on the perceived audiovisual quality, and that this impact depends on the audio degradation type. This is especially true for content C, for both SD and HD, for which regression coefficients for the terms containing $Icod_A$ ($c_{ac}, c_{avcc}, c_{avct}$) are all non-significant, in contrast to the regression coefficients of the terms containing $Itra_A$ ($c_{at}, c_{avtt}, c_{avtc}$).

In addition, c_{vc} is significantly different from c_{vt} for the content C (soccer) of the SD model, highlighting the importance of the video degradation type for this content on the overall audiovisual quality. Note that it has already been observed in the quality-based model that for SD and content C, $\gamma \neq 0$ in Eq. 6.4. The video-only quality and degradation type seem to play a bigger role for content C than for the other contents. Regarding the coefficients of the multiplicative terms

$(c_{avcc}, c_{avtt}, c_{avtc}, c_{avct})$, they are significantly different for several contents (content C for HD, contents A, C, and E for SD). This confirms that assumption "A" needs to be rejected also when modeling the audiovisual quality per content. All those results are in favor of developing an impairment-factor based model, which in addition takes into account the audiovisual content type.

6.2.5 Model Evaluation

For validating the modeling approaches proposed in this chapter, the impairment-factor- and quality- based models have been evaluated against the audiovisual subjective test data-set used for training the model ($ph3 - SD$ and $ph3 - HD$ or "TVM-training-set") and audiovisual subjective test data-set not used for training the model (*ph2—SD* and *ph2—HD* or "TVM-evaluation-set").

Four model variants are compared for each resolution: the content-blind (Q) and -aware (Qc) quality-based models, and the content-blind (IF) and -aware (IFc) impairment-factor-based models. The content-blind models use the same set of coefficients for all contents (see rows "HD all" and "SD all" in Tables 6.2 and 6.3). The content-aware models use one set of coefficients per content (see rows "HD B to E" and "SD A to E" in Tables 6.2 and 6.3).

6.2.5.1 Performance Indicators

Performance indicators used for validating the modeling approach and used in the following analysis are the Pearson correlation coefficient R, the $RMSE$, and the $RMSE^*$ values (see Sect. 3.5.3). Performance results are summarized in Tables 6.5 and 6.6 for HD and SD respectively, for the content-blind (Q) and -aware (Qc) quality-based models according to Eq. (6.4), and for the content-blind (IF) and -aware (IFc) impairment-factor-based models according to Eq. (6.6). Table 6.5 (resp. 6.6) shows the performance of the HD (resp. SD) audiovisual quality models on the TVM training (R_t, $RMSE_t$, $RMSE_t^*$) and TVM evaluation (R_e, $RMSE_e$, $RMSE_e^*$) data set. The impairment factors and audio- and video- qualities are derived from the subjective ratings.

If a model performs significantly better than the content-blind quality-based model Q, the performance indicators (R_y and $RMSE_y^*$, $y \in \{t, e\}$) are marked in italic; if one of the two impairment-factor based models is performing better than the other, the respective performance values is printed in bold. The significance of the differences between performance indicators is tested following the criteria described in Sect. 2.4.3.

Figures 6.4 and 6.5 show the performance of the content-blind impairment-factor-based model on the TVM evaluation data set for HD and SD, when the impairment factors are derived from the subjective tests. This corresponds to the most valid way of evaluating the audiovisual impairment-factor-based model, since the TVM

Table 6.5 Performance for HD, for the TVM training ($_t$) and TVM evaluation ($_e$) data; Audio and video quality and impairment factor terms are derived from subjective tests

	R_t	$RMSE_t$	$RMSE_t$*	R_e	$RMSE_e$	$RMSE_e$*
Q	0.94	0.68	0.30	0.94	0.71	0.28
Qc	0.94	0.60	*0.21*	0.94	0.64	*0.22*
IF	0.96	0.53	*0.15*	0.95	0.57	***0.14***
IFc	***0.98***	0.39	***0.07***	0.95	0.58	*0.24*

Q: content-blind Q-based model, *Qc*: content-aware Q-based model, *IF*: content-blind IF-based model, *IFc*: content-aware IF-based model; In italic: the respective model performs significantly better than the corresponding basic model *Q*; In bold: Significantly better performing model between *IF* and *IFc*

Table 6.6 Performance for SD

	R_t	$RMSE_t$	$RMSE_t$*	R_e	$RMSE_e$	$RMSE_e$*
Q	0.94	0.56	0.20	0.92	0.65	0.26
Qc	0.95	0.52	*0.18*	0.92	0.64	0.26
IF	0.94	0.55	0.18	0.93	0.58	*0.20*
IFc	0.95	0.48	***0.14***	0.92	0.64	*0.22*

See Table 6.5 for more details

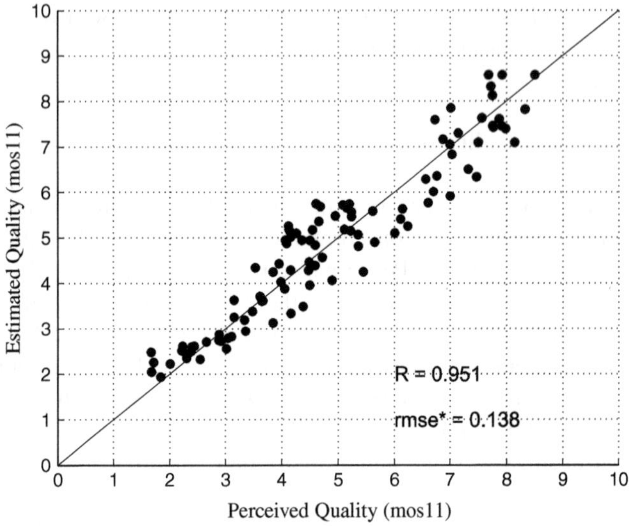

Fig. 6.4 Performance on the 11-point mos11 scale of the content-blind HD impairment-factor-based model on unknown subjective data. Impairment factors are derived from the subjective tests

evaluation data are unknown to the model, and for audiovisual quality prediction the model directly uses the subjective results from the audio- and video-only tests, instead of the audio- and video-only quality models with their possible prediction errors.

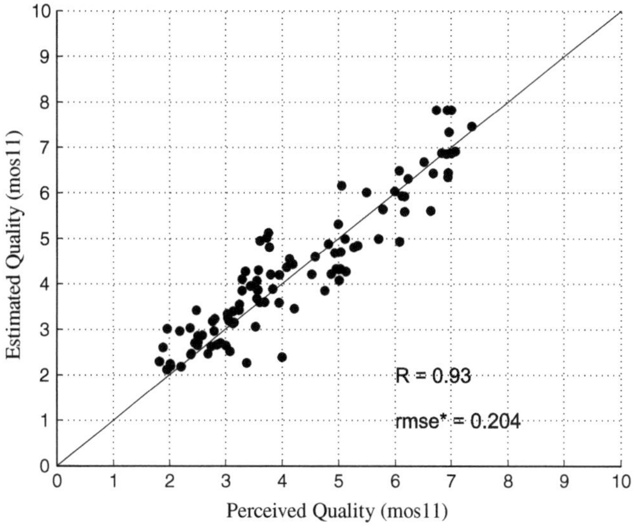

Fig. 6.5 Performance on the 11-point mos11 scale of the content-blind SD impairment-factor-based model on unknown subjective data. Impairment factors are derived from the subjective tests

The model performance comparison will start with general observations for all results, then continue by evaluating the benefit of taking the degradation type into account. In a third stage, the advantage of considering the content type in the model will be analysed.

It can first be observed that all model variants obtain good performance results, especially for HD, where the models always obtain correlations above 0.94, up to 0.98, the *RMSE* is between 0.39 and 0.71, and the *RMSE** is between 0.07 and 0.30 (on the 11-point scale used in the tests). The SD model variants obtain slightly lower performance, with correlation values ranging from 0.86 to 0.95, *RMSE* between 0.48 and 0.65, and *RMSE** is between 0.14 and 0.28. As expected, the *RMSE* and *RMSE** are the most discriminative performance measure between models. For instance, when Q and Q_c obtain the same correlation coefficients for HD on training set ($R_t = 0.94$), their *RMSE* and *RMSE** still differ ($RMSE(Q) = 0.68$ and $RMSE^*(Q) = 0.30$ while $RMSE(Q_c) = 0.60$ and $RMSE^*(Q_c) = 0.21$).

6.2.5.2 Model Evaluation for HD

For HD, the content-blind impairment-factor-based model (*IF*) always performs significantly better then the content-blind quality-based model (Q). Since the *RMSE** takes into account the confidence interval of each sequence, the slightly higher confidence interval values of the evaluation data compared to the training data eases obtaining good performance for all models and therefore increases the difficulty of achieving significant difference between the $RMSE^*$ of different models. Promising result is that the impairment-factor-based model variants *IF* and *IFc* in all cases perform better than the quality-based model variants *Q* and *Qc*. Considering the content in the modeling further improves the performance of the models in all cases except for the evaluation data with the impairment-factor-based model. This may be due to an overtraining of the model. Indeed, even though contents used in the evaluation data set are identical to some of the contents of the training data set, different conditions were used between the two sets. Moreover, the processing chains were different, yielding different perceptual impacts for similar conditions. As a consequence, the evaluation set can be considered to represent a case where different contents between the data sets are used. Since the coefficients are content-specific, the prediction for one content can even be worse than when using the coefficient set of "row all", which were obtained using ratings from several contents.

6.2.5.3 Model Evaluation for SD

The advantage of using the impairment-factor-based approach is not as clear for SD as for HD. However, it can be noticed that the impairment-factor-based model *IF* performs better than the quality-based model *Q* in all cases.

As in the case of HD, considering the influence of the content further improves the performance of the models in all cases, except for the evaluation data with the impairment-factor-based model.

6.2.5.4 Comparison with Models Described in the Literature

The performance of the proposed models is now compared to the performance of models described in the literature. Note that in almost all studies, the models are quality-based, their performance is computed using the training data set, and the audio and video quality terms Q_A and Q_V of the models are fed with the subjective test values. Since validating the models on unknown data is considered to be more suitable, the performance of the best-performing models, i.e., the content-blind impairment-factor-based models for both SD and HD, is shown on the evaluation data set. As a comparison point, the performance results for the content-blind quality model variants is also depicted. This leads to the correlations listed in Table 6.7. Degradation types addressed by each model are also shown, indicating that all other data have been obtained without considering transmission errors.

The content-blind impairment-factor-based model obtains high correlation values, similar to most of the other models. This is even more valuable since this model can be applied to both coding and transmission errors, that is a wider range of degradation types. However, since the models from the literature have been derived for different video formats and applications, comparing correlation coefficients does not allow any conclusions to be drawn on which model performs the best, but rather gives us an indication of relative performance of our model. Moreover, neither the $RMSE^*$,

Table 6.7 Models performance comparison

Model	Degradation type	α	β	γ	ζ	Correlation[a]	Scale[b]
Bellcore93 (ITU-T SG12 1993)	Analog	1.33	0	0	0.11	*0.99*	9
Bellcore94 (ITU-R SG12 1994)	Analog	1.07	0	0	0.11	*0.99*	9
NTIAITU98 (ITU-T SG12 D.038 1998)	Cod.	1.54	0	0	0.12	*0.93*	9
NTIAJones98 (Jones and Atkinson 1998)	Cod.	−0.677	0.217	0.888	0	*0.98*	5
Beerends99 v1 (Beerends and Caluwe 1999)	Analog	1.12	0.007	0.24	0.09	*0.98*	9
Beerends99 v2 (Beerends and Caluwe 1999)	Analog	1.45	0	0	0.11	*0.97*	9
FT98 (Chateau 1998)	Compression	1.76	0	0	0.10	*0.96*	9
BT04 (head and shoulder) (Hands 2004)	Cod.	1.15	0	0	0.17	*0.85*	0–100
BT04 (High motion) (Hands 2004)	Cod.	0.95	0	0.25	0.15	*0.82*	0–100
Ries07 (Fast movement) (Ries et al. 2005)	Cod.	−0.922	0.569	0.506	0.170	*0.91*[c]	5
Ries07 (video call) (Ries et al. 2005)	Cod.	−0.631	0.214	0.012	0.118	*0.90*[c]	5
Winkler06 v1 (Winkler and Faller 2006)	Cod., frame-rate	−1.51	0.456	0.770	0	*0.94*	11
Winkler06 v2 (Winkler and Faller 2006)	Cod., frame-rate	1.98	0	0	0.103	*<0.94*	11
NTT05 (Yamagishi and Hayashi 2005)	Compr., frame-rate, delay	n.a.	n.a.	n.a.	n.a.	*0.94*	5
Impairment-based T-V-M[d] HD	Cod.,trans.	n.a.	n.a.	n.a.	n.a.	0.95	11
Quality-based T-V-M[d] HD	Cod., trans.	28.49	0	0.13	0.006	0.94	11
Impairment-based T-V-M[d] SD	Cod., trans.	n.a.	n.a.	n.a.	n.a.	0.92	11
Quality-based T-V-M[d] SD	Cod., trans.	30.99	0	0	0.006	0.91	11

"Cod.": compression artifacts, "trans.": transmission errors. In italic: model performance evaluation on training data
[a]Correlation coefficients are assumed to be Pearson correlation coefficients
[b]Number of categories
[c]audio and video quality predicted from models
[d]proposed model

nor the*RMSE*, were accessible for most of the models found in the literature. This measure would have been more appropriate for comparing the models, since, as previously mentioned, it is more discriminative, also in the light of the underlying test data.

6.3 Parametric Audiovisual Quality Model

6.3.1 Audiovisual Model

For reflecting the real instrumental assessment situation, the audio and video quality terms and impairment factor terms are now computed from measurements done on the audio and video streams, and not from subjective tests as in the previous section.

Based on the promising results of the previous section, the audiovisual quality model will be developed following the impairment-factor-based approach. A quality-based audiovisual model is also proposed for comparison.

The *quality-based* and *impairment-factor-based* audiovisual quality models presented in this work take as input the audio and video qualities and impairment factors predicted by the models presented in Chaps. 4 and 5. For convenience, the models are reproduced in the following.

The general form of the **audio quality** is:

$$Q_A = Qo_A - Icod_A - Itra_A \tag{6.7}$$

where $Qo_A = 100$, and

$$Icod_A = a_1 \cdot \exp(a_2 \cdot bitrate) + a_3, \tag{6.8}$$

and

$$Itra_A = (b_0 - Icod_A) \cdot \frac{Pfl}{(b_1 \cdot \mu + b_2) + Pfl} \tag{6.9}$$

bitrate is the overall audio bitrate, *Pfl* is the audio frame loss rate (in percent), and μ is the averaged number of consecutively lost audio frames. Values of the coefficients $a_i, i \in [1-3]$ and $b_j, j \in [0-2]$ depend on the employed audio codec (AAC, MP2, MP3, HE-AAC, or AC-3) and are given in Table 4.3.

For estimating the video quality, both the network planning and quality monitoring models will be tested.

The general form of the **video quality model** is as follows:

$$Q_V = Qo_V - Icod_V - Itra_V \tag{6.10}$$

where $Qo_V = 100$.

For the **network planning model**:

$$Icod_V = (a_1 * KeyFrameRate + a_4) \cdot exp(a_2 \cdot BitPerPixel \cdot Res) + a_3, \quad (6.11)$$

with $Res = 2$ for HD1080 and HD720 video, and $Res = 1$ for SD video. *BitPerPixel* is the averaged number of bits per pixel. a_1, a_2, a_3 depend on the employed codec, here $H.264$.

$$Itra_V = (b_0 - Icodn_V) \cdot \frac{Ppl}{Icodn_V \cdot (b_1 \cdot \mu + b_2) + Ppl} \quad (6.12)$$

wherein $Icodn_V = 65$ if $Icod_V > 65$ and $Icodn_V = Icod_V$ otherwise. b_0, b_1, and b_2 depend on the video codec, the packet-loss-concealment (PLC), and on the slicing configuration, i.e., the number of slices per frame.

The regression coefficients values are given in Table 5.3.

For the **quality monitoring model**:

$$Icod_V = (a_1 + offset) \cdot exp(a_2 \cdot BitPerPixel \cdot Res) + a_3 \cdot SceneComp + a_4 \quad (6.13)$$

with $Res = 2$ for HD1080 and HD720 video, $Res = 1$ for SD video. $offset = 10$ for SD formats, and $offset = 0$ for HD formats.

SceneComp is the scene-complexity parameter defined in Eq. (5.5).

In case freezing is used as PLC:

$$Itra_V = b_0 \cdot log(b_1 \cdot dpSEQ \cdot BitPerPixel + 1) \quad (6.14)$$

where *BitPerPixel* is the already defined averaged number of bits per pixel and *dpSEQ* is the overall corruption duration per measurement window, ignoring initial loss on non-reference b frames, normalized over measurement window. *dpSEQ* has already been described in Sect. 5.3.3, Eq. (5.22).

In case slicing is used as PLC:

$$Itra_V = c_0 \cdot log(c_1 \cdot \frac{xwpSEQsz}{Icodn} + 1) \quad (6.15)$$

with $Icodn = 1$ if $Icod < 20$ and $Icodn = 0.1125 \cdot Icod - 1.25$ otherwise. *xwpSEQsz* is the parameter described in Sect. 5.3.2 which captures the spatio-temporal magnitude of packet loss in case slicing is applied as PLC.

The regression coefficients values are given in Table 5.7.

Once the audio and video qualities Q_A and Q_V are computed, they are sent to the quality-based model, using Eq. (6.16):

$$Q_{AV} = \alpha + \beta \cdot Q_A + \gamma \cdot Q_V + \zeta \cdot Q_A \cdot Q_V \quad (6.16)$$

For the impairment-factor-based model, the impairment factors $Icod_A$, $Icod_V$, $Itra_A$, and $Itra_V$ are used instead, and directly inserted into Eq. (6.17), with $Qo_{AV} = 100$.

$$
\begin{aligned}
Q_{AV} = Qo_{AV} \\
& -b \cdot Icod_A - c \cdot Icod_V \\
& -d \cdot Icod_A \cdot Icod_V \\
& -e \cdot Itra_A - f \cdot Itra_V \\
& -g \cdot Itra_A \cdot Itra_V \\
& -h \cdot Icod_V \cdot Itra_A - i \cdot Icod_A \cdot Itra_V
\end{aligned}
\tag{6.17}
$$

Since the audiovisual (AV) quality models have been developed in the context of the P.NAMS competition, and since the P.NAMS training database set contains a sufficiently high number of audiovisual databases, the AV quality models are trained on the P.NAMS training database sets (*tr*12, *tr*13, and *tr*14).

Four models are ultimately trained:

- Q_{AV}^{NP}: a *quality*-based AV model which takes as input the *quality terms* estimated by the audio and *network planning* video quality models.
- IF_{AV}^{NP}: an *impairment-factor*-based AV model which takes as input the *impairment-factor* terms estimated by the audio and *network planning* video quality models.
- Q_{AV}^{QM}: a *quality*-based AV model which takes as input the *quality terms* estimated by the audio and *quality monitoring* video quality models.
- IF_{AV}^{QM}: an *impairment-factor*-based AV model which takes as input the *impairment-factor* terms estimated by the audio and *quality monitoring* video quality models.

As for the audio and video quality models, and since the P.NAMS validation set was not available at the time of model development, a cross-validation procedure is run on the training set for preventing overtraining. The models are eventually evaluated against unknown data using the P.NAMS validation set (*vl*21, *vl*22, *vl*23 and *vl*24).

6.3.2 Model Evaluation

6.3.2.1 Cross-Validation

As for the audio and video quality models, the cross-validation (CV) is done on the training data (*tr*12, *tr*13, and *tr*14), and at source content level: at each iteration of the CV, the model is trained on six contents and evaluated against the remaining two contents. This results in 28 permutations of the content, and therefore 28 iterations of the CV. The CV is conducted for both the "Q-based" and "IF-based" models. The AV models do not distinguish between the SD and HD formats so that a single set of coefficients is valid for all SD and HD formats.

Table 6.8 Regression coefficients (from Eqs. (6.16) and (6.17)) kept for modeling and for each model variant following the cross-validation procedure

Model	Coefficients
Q_{AV}^{NP}	α and ζ
Q_{AV}^{QM}	α, γ and ζ
IF_{AV}^{NP}	b, c, d, f and i
IF_{AV}^{QM}	b, c, e, f, g, h and i

A model is considered overtrained when at least one of its coefficients (for instance coefficient β in the "Q-based" model) has negative and positive values for different iteration of the CV. The coefficient(s) is (are) then said "unstable". In that case, the CV is rerun on the model without the parameter corresponding to the most unstable coefficient (in our example, the cross-validation would be re-run using $Q_{AV} = \alpha + \gamma \cdot Q_V + \zeta \cdot Q_A \cdot Q_V$ as "Q-based" model) and performance indicators are re-computed. If the performance indicators are unchanged, the simplified model is kept. This procedure is conducted till only stable regression coefficients are obtained without degrading the performance indicators. The resulting model is considered as non-overtrained. Following this procedure, it was decided to keep, for the four model variants, the coefficients shown in Table 6.8. The corresponding performance indicators (Pearson correlation R, $RMSE$, and $RMSE^*$) are shown in Table 6.9.

As for the TVM database, the interaction term $Q_A \cdot Q_V$ (coeff: ζ) is predominant in both "Q-based" models, while the audio-quality-only term (coeff: β) is disappearing. The video-quality-only term Q_V (coeff: γ) is part of the model only when using the quality monitoring video quality model. This may indicate that the network planning model is introducing prediction errors which are destabilizing the audiovisual quality model, while the quality monitoring quality model is accurate enough for capturing the predominance of both the video quality and interaction between audio and video quality over the overall audiovisual quality. Similar observation can be made on the "IF-based" model, for which more stable coefficients are found when using the quality monitoring video quality model (IF_{AV}^{QM}).

Interestingly, the interaction terms $Icod_A \cdot Icod_V$ (coeff: d) disappeared from the original Eq. (6.17) for IF_{AV}^{QM}. This term reflects the interaction between video compression artifacts, captured by the IF $Icod_V$, and audio compression artifacts, captured by $Icod_A$. More precisely, the absence of this term in the equation indicates that in the case of vidco and audio compression artifacts, the impact of video degradation on the overall audiovisual quality does not depend on the audio degradation level, and vice versa ($d = 0$). This was not the case in TVM data, most probably due to a different balance between the audio and video compression artifact ranges.

Table 6.9 Cross-validation performance results of the four variants of the AV model

Model		R	RMSE	*RMSE**
Q_{AV}^{NP}	AVG	0.81	0.64	0.43
Q_{AV}^{NP}	STD	0.02	0.07	0.07
Q_{AV}^{NP}	MIN	0.77	0.53	0.32
Q_{AV}^{NP}	MAX	0.86	0.77	0.57
IF_{AV}^{NP}	AVG	0.82	0.62	0.42
IF_{AV}^{NP}	STD	0.03	0.07	0.08
IF_{AV}^{NP}	MIN	0.76	0.49	0.28
IF_{AV}^{NP}	MAX	0.88	0.75	0.59
Q_{AV}^{QM}	AVG	0.86	0.55	0.37
Q_{AV}^{QM}	STD	0.03	0.09	0.09
Q_{AV}^{QM}	MIN	0.80	0.43	0.23
Q_{AV}^{QM}	MAX	0.91	0.74	0.57
IF_{AV}^{QM}	AVG	0.86	0.55	0.37
IF_{AV}^{QM}	STD	0.03	0.09	0.09
IF_{AV}^{QM}	MIN	0.80	0.42	0.23
IF_{AV}^{QM}	MAX	0.91	0.74	0.57

AVG average over all CV runs; *STD* Standard Deviation; *MIN* Minimum; *MAX* Maximum

6.3.2.2 Model Evaluation

The four variants of the audiovisual quality models are now trained on all P.NAMS training data ($tr12$, $tr13$, and $tr14$), still using Eqs. (6.16) and (6.17), but only keeping the coefficients of Table 6.8 (the other coefficients are set to zero).

The training results in the model regression coefficients shown in Table 6.10 for the four variants of the model. Regression coefficients are valid for both SD and HD.

The coefficients of the IF_{AV}^{QM} model were compared to the TVM coefficients of Table 6.3. Although they were not trained on the same databases, certain tendencies are common to both set of coefficients, especially for HD: the video degradations are dominating over the audio degradations ($(c, f) > (b, e)$), and the quality impact due to the audio transmission errors is perceptually stronger than for audio compressions in the audiovisual context ($b < e$).

The models are then evaluated against the validation data, using the P.NAMS validation set ($vl21$, $vl22$, $vl23$, and $vl24$, i.e., 960 audiovisual files).

As mentioned in Sect. 3.5.3, the quality scores predicted by the audiovisual quality models were first transformed to the 5-point MOS scale. Then a first-order polynomial mapping was applied per validation database on the transformed output and using the

Table 6.10 Regression coefficients [from Eqs. (6.16) and (6.17)] for the four variants of the model

Model	α	γ	ζ					
Q_{AV}^{NP}	29.07	0	0.007					
Q_{AV}^{QM}	8.99	0.42	0.006					
Model	*b*	*c*	*d*	*e*	*f*	*g*	*h*	*i*
IF_{AV}^{NP}	0.55	0.60	−0.004	0	1.12	0	0	−0.012
IF_{AV}^{QM}	0.35	0.82	0	0.60	0.99	−0.007	−0.005	−0.008

Table 6.11 Performance results on unknown data of the four variants of the AV model

Model	R	RMSE	*RMSE**
Q_{AV}^{NP}	0.84	0.55	0.36
IF_{AV}^{NP}	0.86	**0.52**	0.34
Q_{AV}^{QM}	0.92	*0.42*	0.24
IF_{AV}^{QM}	0.92	*0.42*	0.24

subjective ratings (averaged over all subjects) of each validation database as target values.

The same performance indicators as for the cross-validation are computed per database. For each performance indicator, the per-database values are then averaged over all validation databases. Results are listed in Table 6.11 for the four variants of the AV model. The significance of the difference between the performance indicators is tested using the criteria described in Sect. 2.4.3. If the "Q-based" and "IF-based" Quality Monitoring models perform significantly better than the respective "Q-" and "IF-" based network planning models, they are marked in italic. If the "IF-based" models perform significantly better than the "Q-based" models, they are marked in bold. Performance results of the IF_{AV}^{QM} model on unknown data are in addition plotted in Fig. 6.6.

As expected, both variants of the audiovisual quality model using the quality monitoring video quality model, Q_{AV}^{QM} and IF_{AV}^{QM}, are performing significantly better than the variants based on the network planning video quality model. Indeed, taking the example of the comparison of the "IF-based" models, the q value of Eq. (2.13) becomes $q = \frac{rmse_h^2}{rmse_l^2} = \frac{0.52^2}{0.42^2} = 1.53$, which is bigger than the tabulated F value: $F(0.95, 920-2, 920-2) = 1.11$.

In addition, the "IF-based" model is performing significantly better than the "Q-based" model for network planning.

However, it is performing equally good but not better than the "Q-based" model in the case of quality monitoring.

Considering that (a) the "IF-based" model is performing better on TVM-data (see Sect. 6.2.5), and (b) the "IF-based" model is performing better than the "Q-based"

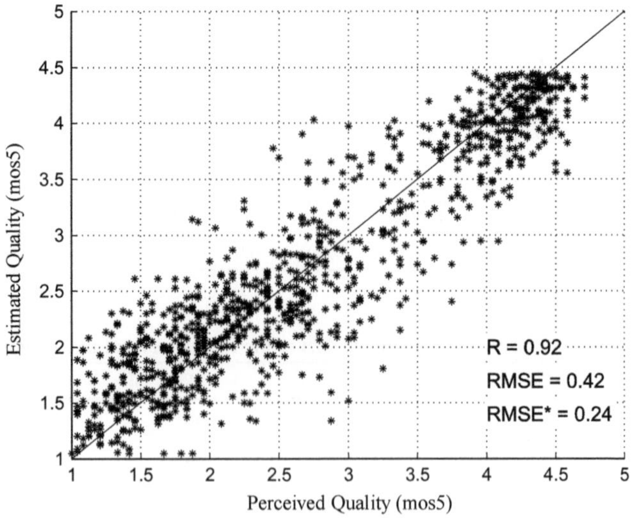

Fig. 6.6 Performance on the 5-point (*mos*5) scale and on unknown data of the AV quality model variant IF_{AV}^{QM}

model for network planning (see above), this result is rather disappointing. After further investigation, it has been observed that Q_{AV}^{QM} and IF_{AV}^{QM} output the same quality scores on the validation database set.

These results find an explanation in the nature of the validation database set. Indeed, the variation of audio quality for both audio compression and audio transmission errors is much smaller in the validation set than in the training set. As an example, in the no-loss case, $Icod_A$ is taking values up to 80 in the training set, while its maximum value in the validation set is 55. Similarly, the maximum value for $Itra_A$ is 50 in the training set, while it is only 40 in the validation set. At this stage, an important observations made in the previous sections should be reminded: the impact on Q_{AV} of video compression artifact depends on the audio degradation type [$d \neq h$ in Eq. (6.17)], and this impact depends as well on the amount of audio packet loss. The users' attention is attracted more to audio only in the case of transient audio degradations such as audio frame loss. This phenomena is well captured by the "IF-based" model, having $d \neq h$, but if there is almost no audio frame loss, at least compared to the amount of video loss, the advantage of using the "IF-based" model over the "Q-based" model is fading.

For the ITU P.NAMS competition, a combined "Q-based" and "IF-based" model was produced, using Eq. (6.18):

$$Q_{AV} = w_1 \cdot Q_{AV}^{QM} + w_2 \cdot IF_{AV}^{QM}, \tag{6.18}$$

Best performing model was obtained for $w_1 = 0.3$ and $w_2 = 0.7$. This model won the ITU-T P.NAMS competition and is part of the ITU-T P.1201 standard (ITU-T Recommendation P.1201 2012, ITU-T Recommendation P.1201.2 2012).

6.4 Summary

Based on the results of numerous audiovisual subjective tests, the quality impact due to audio and video on the audiovisual quality in the context of IPTV services has been analysed. Results show that video quality is dominating the audiovisual quality over the audio quality. However, the most dominant factor of audiovisual quality is the interaction between audio and video qualities, highlighting that the quality impact due to video on the audiovisual quality depends on the audio quality level, and vice versa.

It has also been shown that these results depend on the audiovisual content type and on the degradation type (compression vs. transmission errors). The audio saliency increases with contents such as music video, especially compared to contents such as soccer. It has also been observed that the quality impact due to video on the audiovisual quality increases with the video format, which is an intuitive result. One remarkable observed behavior is that in an audiovisual test, the subjects seem to focus more on video, as in a video-only test, while the audio is only subconsciously attended to. With their main attention on the video, the subjects pay almost similar attention to stationary degradations such as compression artifacts as to more time-varying degradations such as transmission errors, just as in a video-only test. The users' attention is attracted more to the audio only in the case of transient audio degradations such as audio frame loss.

Following these observations, two modeling approaches were tested, one based on audio and video quality terms—the "Q-based" model –, and one based on the audio and video impairment factors (IF) terms linked to compression and transmission errors—the "IF-based" model. As a first step, these two models were developed using audio and video subjective ratings as inputs to the model. Performance results confirmed the validity of the IF-based approach.

Following this positive result, both modeling approaches were tested in the network planning and service (quality monitoring) cases, yielding four models. The audio quality model of Chap. 4 was used as input to all audiovisual quality models. In the case of network planning, the network planning packet-based video quality model of Chap. 5 was used, while the quality monitoring frame-based video quality models of the same chapter was used as input to the audiovisual quality model in the case of quality monitoring. All models output an estimate of the audiovisual quality on the 100-point model-scale, or on the 5-point-MOS-scale after transformation of the outputted scores.

All models were first validated following a cross-validation procedure, and then, once available, on unknown data. With a Pearson correlation coefficient of $R = 0.86$ with subjective tests results and a *RMSE* of 0.52, the "IF-based" model obtained better performing results than the "Q-based" model in the case of network planning on unknown data. The quality monitoring models perform equally well. They both outperform the network planning model, with $R = 0.92$ and $RMSE = 0.42$ on the same unknown data. The unknown databases were unfortunately not sufficiently discriminative in terms of audio quality ranges for differentiating the "Q-based" from the "IF-based" model in the case of quality monitoring. A combined "Q-based" and "IF-based" audiovisual quality model was sent to the ITU-T P.NAMS competition. This combined model won the competition and is standardized under the ITU-T P.1201.2 Recommendation (ITU-T Recommendation P.1201.2 2012) [see also (Garcia et al. 2011; Garcia et al. 2013)].

Chapter 7
Conclusion and Outlook

7.1 Conclusion and Discussion

The two main outcomes of this work are a set of subjective test databases—along with their analyses—produced in the context of IPTV applications, and a standardized parametric audiovisual quality model which estimates the audio, video, and audiovisual quality of encrypted multimedia streams for IPTV services. This work also provides a detailed description of the steps to be completed for developing and evaluating the quality model. It clarifies too how the model can be embedded in a general QoE framework.

Another interesting output of this work is the validation of an anchor-based method for merging subjective test results from different databases, which is an important step to be completed before applying any data modeling.

The analysis of the subjective test results provided a better understanding of the origin of the typical IPTV audio and video degradations, as well as of their respective impact on the perceived quality.

The model addresses the IP/RTP/TS packetization scheme and works on PES- and TS-level encrypted audiovisual streams. It would also work on other transport protocols containing TS, such as IP/UDP/TS, but since the packet headers would differ, the data acquisition and parameter aggregation described in the ITU-T P.1201.2 standard would need to be adapted.

Covered degradations are compression artifacts, such as blockiness and blurriness, as well as transmission errors yielding freezing or slicing degradations. For slicing, the model was ultimately validated on data encoded with either 1 slice per frame or 1 slice per Macro-block row. The model is however robust to other slicing configurations since the number of slices per frame is included in the model input parameters computation. Covered codecs are H.264 (Main and High Profile) for video, and AAC-LC, HE-AAC, MP2, MP3 and AC3 for audio. Several GOP structures were addressed such as IbbP, IPPP, and IbBbP (hierarchical coding), as well as different GOP lengths. Both random and bursty losses are covered by the model. The model has been validated on SD (PAL and NTSC), HD720 and HD1080 resolutions,

M.-N. Garcia, *Parametric Packet-Based Audiovisual Quality Model for IPTV Services*, T-Labs Series in Telecommunication Services, DOI: 10.1007/978-3-319-04855-0_7,

on interlaced and progressive format, and on different frame rates (25 fps, 30 fps, 50 fps and 60 fps). However, by including the resolution (number of pixels per frame) and the frame rate into the computation of the model input parameters, the proposed model is robust to other video formats and frame rates without re-training the model.

It should be highlighted that the originally purely additive impairment-factor modeling approach inspired by Allnatt's work—which assumes that certain kind of impairment factors may be considered additive on an appropriate (perceptual) quality rating scale—evolved during the modeling towards a mixed additive-multiplicative approach. The proposed models are indeed not only adding the single impairment factor terms corresponding to the quality impact due to compression artifacts and transmission errors, but the interaction terms as well. This is highlighting that the quality impact due to a given degradation type depends on the level of the other(s) degradation type(s). This deviation from a purely additive model to a mixed model may also indicate that the selected impairment factors do not perfectly capture the quality dimensions underlying the perceived audio, video, and audiovisual quality. Indeed, it was assumed that two quality dimensions were underlying the perceived quality in the tests, one dimensions being linked to compression-related artifacts, the other dimension being linked to ransmission errors. A multidimensional analysis, or Semantic Differential analysis, should be conducted for revealing if this assumption was true and, if not, for helping finding more appropriate impairment factors.

In any case, and despite this deviation from the purely additive modeling approach, all models obtained high performance results on unknown data and in the context of IPTV services.

The model has been standardized by the ITU under the ITU-T Recommendation P.1201 (ITU-T Recommendation P.1201 2012) and its extension for higher resolution application ITU-T P.1201.2 (ITU-T Recommendation P.1201.2 2012) [see also (Garcia et al. 2011, 2013a, b)].

It is now worth investigating which steps should be undertaken for implementing the model in the context of service (quality) monitoring, and for using it in order to estimate the QoE.

7.2 Towards Quality Monitoring

Since the data acquisition from the encrypted bitstream was not part of the present work, the model implementation is not described in this book. The data acquisition and parameter extraction were however parts of the P.NAMS competition, and are fully described in the ITU P.1201.2 document (ITU-T Recommendation P.1201.2 2012).

The implemented models acquire data on measurement window of 10 s duration—a small deviation from this duration would not hurt the accuracy of the model predictions—and therefore output estimates of the perceived quality every 10 s. These estimates may be used for computing the so-called Key Quality Indicators (KQIs). As an example, the impairment frequency tolerance threshold (Raake 2012) indicates from which impairment frequency the user of the service may start complaining. The DSL TR-126 report (DSL Forum 2006) states that this frequency should not be higher than one impairment per hour in the case of SD video formats, and one impairment every four hours in the case of HD formats. This latter threshold has been reduced to one impairment every one hour in Staelens et al.'s work (Staelens et al. 2010). It should now be decided from which quality decrease the monitoring system should report an impairment. Typical 95 % confidence interval in the conducted subjective tests with approx. 24 subjects per test is 0.5 on the 5-point MOS scale. As a consequence, it is proposed to report an impairment for an event yielding a quality decrease of 0.5 MOS.

7.3 Towards QoE

These KQIs are still not QoE. They are not sufficient for knowing if a customer is satisfied with the service she/he subscribed to. To find the link between the quality scores provided by the quality models and user satisfaction, several points have to be addressed:

- The dimensions underlying QoE for a given application should be identified. For IPTV, these dimensions may be for instance the perceived quality, accessibility/stability and usability/joy of use. Subjective tests are required for identifying these QoE dimensions. Multidimensional analysis and interview-based methods are examples of test methods which could be applied in that respect.
- QoE-models should be made ecologically valid, user- and context-dependent. One requirement is bringing subjective testing to the field. In addition, new subjective tests should be designed for addressing long audiovisual sequences (as with the SSCQE of the ITU-R BT-500 Rec.), but with more appropriate tasks so that the subjects focus on the content and not on the quality only [in the latter case, subjects are becoming too sensitive to impairments, see (Staelens et al. 2010)]. For making QoE-models user-dependent, the most QoE-relevant criteria (personality? degree of expertise? Demographic data?) for characterizing the users should be identified.

A bit of investigation work has been conducted on the integration of user-related information into the model, if it is found useful to build a user-dependent model. It should be first investigated at which time window the user-dependency is happening. Indeed, user factors may impact the user experience at four different levels:

1. On the short-term (e.g., 10 s) audio and video quality estimates Q_A and Q_V. This dependency would for instance reflect that different groups of users have different sensibilities to audio and/or video degradations. The estimated Q_A and Q_V values would then differ between the user groups.
2. On the short-term audiovisual quality estimates Q_{AV}: a group of users may be more sensitive to audio degradations than video degradations in the audiovisual context, while another group of users may be more sensitive to video degradations.

In the case of the quality-based approach, case for which the audiovisual quality is computed from the estimated audio and video qualities, as shown in Eq. (7.1), the α, β, γ and ζ coefficients would then vary from one user group to another one.

$$Q_{AV} = \alpha + \beta \cdot Q_A + \gamma \cdot Q_V + \zeta \cdot Q_A \cdot Q_V \tag{7.1}$$

3. On the per-session estimates of audiovisual quality, where a session has for instance, and in the case of IPTV, the duration of a movie or of a TV-program, e.g., approx. $1h30$. The per-session audiovisual quality would be estimated as a weighted function of the short-term audiovisual quality scores. These weights, and therefore weighting functions, would depend on the user group.
4. At KQIs level: If the selected KQI is the number of detected impairments per hour, the threshold for determining if there is an impairment and/or of the impairment frequency tolerance could be adjusted depending on the user group.

Note that the user groups may differ between these four time windows. The output of the user classification—and therefore the user groups—may also depend on the context, i.e., whether the tests for determining these groups are conducted in the laboratory or in the field. The task given to the subjects will also most probably impact the user classification and should therefore be carefully selected.

7.4 Model Extensions

The model proposed in this work is already foreseen to be extended to "Portable TV", which is an IP-based application in which the user is able to watch TV-programs whenever and wherever she/he wants to, including outside her/his home. The model should in particular be adapted to technological advancement such as HTTP-based adaptive streaming, e.g., Dynamic Adaptive Streaming over HTTP (mpeg-DASH) standard, and to High Efficiency Video Coding (HEVC) coding, the successor of H.264. This includes adapting the model to new types of degradations, such as initial loading and stalling. Other quality aspects should be considered such as the potential quality variation between one screen to another one, or due to different user expectations and sensibilities depending on the display or on the location (home vs. outside).

It is unclear how the future TV will be. It will be most probably personalized, with personal menus and program suggestions. Future TV will also be interactive with for instance the possibility of chatting at the same time a TV-program is watched, or by virtually participating in a TV-show or concert. A completely new paradigm is required for assessing the quality of personalized and interactive TV experience. It is therefore too early for knowing how quality models could be used in that context. Instead, it is a great opportunity for including the increasing knowledge on QoE into the conceptual phase of an application, and therefore impacting the design of the application itself.

Appendix A

A.1 5-Point MOS Scale to 100-Point Model-Scale

This section provides the C-code used for transforming the quality scores obtained on the 5-point MOS scale to the quality scores to be expressed on the 100-point scale, as defined in the Appendix A of ITU-T Recommendation G.107 (2005).

If MOS and Q are quality scores expressed on the 5-point-MOS and 100-point scales respectively, then:

function $Q = Q from\, MOS(MOS)$
set $x = 18566 - 6750 \cdot MOS$;
if $MOS > 4.5$, $MOS = 4.5$; end
if $x < 0$,
$h = \frac{1}{3} \cdot (\pi - \arctan(\frac{15 \cdot \sqrt{-903{,}522 + 1{,}113{,}960 \cdot MOS - 202{,}500 \cdot MOS^2}}{6{,}750 \cdot MOS - 18{,}566}))$;
else
$h = \frac{1}{3} \cdot \arctan(\frac{15 \cdot \sqrt{-903{,}522 + 1{,}113{,}960 \cdot MOS - 202{,}500 \cdot MOS^2}}{18{,}566 - 6{,}750 \cdot MOS})$;
end
$Q = \frac{20}{3} \cdot (8 - \sqrt{226} \cdot \cos(h + \frac{\pi}{3}))$;

A.2 100-Point Model-Scale to 5-Point MOS Scale

This section provides the C-code used for transforming the quality scores obtained on the 100-point model-scale to the quality scores to be expressed on the 5-point MOS scale.

function $MOS = MOS from\, Q(Q)$
set $MOS_{MAX} = 4.9$;
set $MOS_{MIN} = 1.05$;

M.-N. Garcia, *Parametric Packet-Based Audiovisual Quality Model for IPTV Services*, T-Labs Series in Telecommunication Services, DOI: 10.1007/978-3-319-04855-0,

if $Q > 0 \,\&\, Q < 100$,
$MOS = MOS_{MIN} + \frac{MOS_{MAX} - MOS_{MIN}}{100} \cdot Q + Q \cdot (Q - 60) \cdot (100 - Q) \cdot 7 \cdot 10^{-6}$;
elseif $Q \geq 100$,
$MOS = MOS_{MAX}$;
else
$MOS = MOS_{MIN}$;
end

Appendix B

The detailed test plans of the subjective tests entirely processed at T-Labs are provided in this Appendix. The list of detailed test plans is summarized per database set and per modality in Table B.1.

The following abbreviations are used in the test plans:

- ID: IDentification Number of test condition,
- uncomp: Uncompressed (No degradation),
- ACOD: Audio CODec,
- VCOD: Video CODec,
- ACBR: Constant Audio BitRate,
- VCBR: Constant Video BitRate,
- KFR: Key-Frame-Rate (Number of I-frames per second),
- GOP: GOP structure, e.g., *M*3*N*50: 1 I-frame every 50 frames, 2 non-reference B-frames (b) between P/I-frames (IbbP),
- Nsl: Number of slices per frame or per MacroBlock Row,
- Ppl: Percentage (RTP) packet-loss,
- Pfl: Percentage audio frame-loss,
- PLC: Packet-Loss-Concealment,
- PLD: Packet-Loss-Distribution,
- mu: Average burst length.

Abbreviations specific to a single database are expended in the caption of this database (Tables B.2, B.3, B.4, B.5, B.6, B.7, B.8, B.9, B.10, B.11, B.12, B.13, B.14, B.15, B.16, B.17, B.18, B.19 and B.20)

M.-N. Garcia, *Parametric Packet-Based Audiovisual Quality Model for IPTV Services*, T-Labs Series in Telecommunication Services, DOI: 10.1007/978-3-319-04855-0,

Table B.1 List of T-Labs test plans for the TVM and P.NAMS training databases

ID	Modality	Set	Table ID
$ph2_A$	Audio	TVM	Table B.2
$ph3_A$	Audio	TVM	Table B.3
ref-HD	Video (HD1080)	TVM	Table B.4
test1-HD	Video (HD1080)	TVM	Table B.5
test2-HD	Video (HD1080)	TVM	Table B.5
test3-HD	Video (HD1080)	TVM	Table B.6
test4-HD	Video (HD1080)	TVM	Table B.7
test5-HD	Video (HD1080)	TVM	Table B.8
ref-SD	Video (SD PAL)	TVM	Table B.9
test1-SD	Video (SD PAL)	TVM	Table B.10
test2-SD	Video (SD PAL)	TVM	Table B.10
test3-SD	Video (SD PAL)	TVM	Table B.11
test4-SD	Video (SD PAL)	TVM	Table B.12
test5-SD	Video (SD PAL)	TVM	Table B.13
$ph2_{AV}^{HD}$	Audiovisual (HD1080)	TVM	Table B.14
$ph3_{AV}^{HD}$	Audiovisual (HD1080)	TVM	Table B.15
$ph2_{AV}^{SD}$	Audiovisual (SD PAL)	TVM	Table B.16
$ph3_{AV}^{SD}$	Audiovisual (SD PAL)	TVM	Table B.17
$tr16$	Audio	P.NAMS training	Table B.18
$tr10$	Video (HD720)	P.NAMS training	Table B.19
$tr13$	Audiovisual (HD720)	P.NAMS training	Table B.20

SD PAL: 576p24 and 576i25; HD1080: 1080p24 and 1080i25; HD720: 720p50

Table B.2 $ph2_A$ test plan, audio

ID	ACOD	ACBR (kbps)	Pfl (%)	PLD
hrc01	Uncomp.	–	–	
hrc02	Band limitation 3.5 kHz	–	–	
hrc03	Band limitation 7 kHz	–	–	
hrc04	Distortion MNRU 12 dB	–	–	
hrc05	MP2	96	0	
hrc06	MP2	128	0	
hrc07	MP2	192	0	
hrc08	MP3	64	0	
hrc09	MP3	96	0	
hrc10	MP3	128	0	
hrc11	AAC-LC	48	0	
hrc12	AAC-LC	64	0	
hrc13	AAC-LC	96	0	
hrc14	HE-AACv2	32	0	
hrc15	HE-AACv2	48	0	
hrc16	HE-AACv2	64	0	

(continued)

Table B.2 (continued)

ID	ACOD	ACBR (kbps)	Pfl (%)	PLD
hrc17	MP2	96	1	Random
hrc18	MP2	96	4	Random
hrc19	MP2	96	8	Random
hrc20	MP2	128	1	Random
hrc21	MP2	128	4	Random
hrc22	MP2	128	8	Random
hrc23	MP2	192	1	Random
hrc24	MP2	192	4	Random
hrc25	MP2	192	8	Random
hrc26	MP3	64	1	Random
hrc27	MP3	64	2	Random
hrc28	MP3	64	4	Random
hrc29	MP3	96	1	Random
hrc30	MP3	96	2	Random
hrc31	MP3	96	4	Random
hrc32	MP3	128	1	Random
hrc33	MP3	128	2	Random
hrc34	MP3	128	4	Random
hrc35	AAC-LC	48	1	Random
hrc36	AAC-LC	48	4	Random
hrc37	AAC-LC	48	8	Random
hrc38	AAC-LC	64	1	Random
hrc39	AAC-LC	64	4	Random
hrc44	HE-AACv2	32	1	Random
hrc45	HE-AACv2	32	4	Random
hrc46	HE-AACv2	32	8	Random
hrc47	HE-AAC v2 no concealment	32	4	Random
hrc48	HE-AACv2	64	1	Random
hrc49	HE-AACv2	64	4	Random
hrc50	HE-AACv2	64	8	Random

Table B.3 $ph3_A$ test plan, audio

ID	ACOD	ACBR (kbps)	Pfl (%)	PLD	PLC
hrc01	Uncomp.				
hrc02	Band limitation 3.5 kHz				
hrc03	Band limitation 7 kHz				
hrc04	Distortion MNRU 12 dB				
hrc05	AAC-LC	48			
hrc06	MP2	192	8	Random	
hrc07	MP2	48			
hrc08	MP2	64			
hrc09	MP3	32			
hrc10	MP3	48			
hrc11	AAC-LC	24			
hrc12	AAC-LC	32			
hrc13	HE-AAC	16			
hrc14	HE-AAC	24			
hrc15	AAC-LC	96			
hrc16	MP2	192			
hrc17	MP2	192	1	Bursty (bfrl = 1)	n
hrc18	MP2	192	1	Bursty (bfrl = 2)	n
hrc19	MP2	192	1	Bursty (bfrl = 4)	n
hrc20	MP2	192	1	Bursty (bfrl = 6)	n
hrc21	MP2	192	4	Bursty (bfrl = 1)	n
hrc22	MP2	192	4	Bursty (bfrl = 2)	n
hrc23	MP2	192	4	Bursty (bfrl = 4)	n
hrc24	MP2	192	4	Bursty (bfrl = 6)	n
hrc25	MP2	192	8	Bursty (bfrl = 2)	n
hrc26	MP2	192	8	Bursty (bfrl = 4)	n
hrc27	MP2	192	8	Bursty (bfrl = 6)	n
hrc28	MP2	96	1	Bursty (bfrl = 2)	n
hrc29	MP2	96	1	Bursty (bfrl = 6)	n
hrc30	MP2	96	4	Bursty (bfrl = 2)	n
hrc31	MP2	96	4	Bursty (bfrl = 6)	n
hrc32	MP2	96	8	Bursty (bfrl = 2)	n
hrc33	MP2	96	8	Bursty (bfrl = 4)	n
hrc34	MP2	96	8	Bursty (bfrl = 6)	n
hrc35	AAC-LC	96	1	Bursty (bfrl = 1)	y
hrc36	AAC-LC	96	4	Bursty (bfrl = 1)	y
hrc37	AAC-LC	96	8	Brsty (bfrl = 1)	y
hrc38	AAC-LC	96	1	Bursty (bfrl = 1)	n
hrc39	AAC-LC	96	4	Bursty (bfrl = 1)	n
hrc40	AAC-LC	96	8	Bursty (bfrl = 1)	n

For *PLC*: *n*: no; *y*: yes. *bfrl*: burst length at audio frame level

Table B.4 *Ref* test plan, video, HD1080. *dur.*: duration

ID	VCOD	KFR (Hz)	VCBR	Ppl	PLD	PLC
hrc01	Uncomp.					
hrc02	H.264	1	32			
hrc03	H.264	1	16			
hrc04	H.264	1	8			
hrc05	H.264	1	4			
hrc06	H.264	1	2			
hrc07	MPEG2	1	64			
hrc08	MPEG2	1	32			
hrc09	MPEG2	1	16			
hrc10	MPEG2	1	8			
hrc11	MPEG2	1	4			
hrc12	Uncomp.			5 % dur.		Freezing
hrc13	Uncomp.			10 % dur.		Freezing
hrc14	Uncomp.			20 % dur.		Freezing
hrc15	Uncomp.			1 out of 2 frames		Freezing
hrc16	Uncomp.			1 out of 5 frames		Freezing
hrc17	Uncomp.				Slight blurring	
hrc18	Uncomp.				Medium blurring	
hrc19	Uncomp.				Strong blurring	
hrc20	H.264	10	16			
hrc21	H.264	25	16			
hrc22	H.264	1	16	1		Slicing
hrc23	H.264	1	16	2		Slicing
hrc24	H.264	1	16	4		Slicing

Table B.5 Test plan used for *Test*1 and *Test*2, video, HD1080

ID	VCOD	KFR (Hz)	VCBR	Ppl	PLD	PLC
hrc01	MPEG2	1	32	0	Random	
hrc02	H.264	1	16	0	Random	
hrc03	H.264	1	8	0	Random	
hrc04	H.264	1	4	0	Random	
hrc05	MPEG2	1	32	0.5	Random	Slicing
hrc06	MPEG2	1	32	1	Random	Slicing
hrc07	MPEG2	1	32	2	Random	Slicing
hrc08	H.264	1	16	0.02	Random	Freezing
hrc09	H.264	1	16	0.06	Random	Freezing
hrc10	H.264	1	16	0.125	Random	Freezing
hrc11	H.264	1	16	0.25	Random	Freezing
hrc12	H.264	1	16	0.125	Random	Slicing
hrc13	H.264	1	16	0.25	Random	Slicing
hrc14	H.264	1	16	0.5	Random	Slicing
hrc15	H.264	1	16	1	Random	Slicing
hrc16	H.264	1	16	2	Random	Slicing
hrc17	H.264	1	16	4	Random	Slicing
hrc18	H.264	1	8	0.02	Random	Freezing
hrc19	H.264	1	8	0.06	Random	Freezing
hrc20	H.264	1	8	0.125	Random	Freezing
hrc21	H.264	1	8	0.125	Random	Slicing
hrc22	H.264	1	8	0.5	Random	Slicing
hrc23	H.264	1	8	2	Random	Slicing
hrc24	H.264	1	4	0.02	Random	Freezing
hrc25	H.264	1	4	0.06	Random	Freezing
hrc26	H.264	1	4	0.125	Random	Freezing
hrc27	H.264	1	4	0.125	Random	Slicing
hrc28	H.264	1	4	0.25	Random	Slicing
hrc29	H.264	1	4	0.5	Random	Slicing
hrc30	H.264	1	4	1	Random	Slicing
hrc31	H.264	1	4	2	Random	Slicing
hrc32	Uncomp.					
hrc33	H.264	1	2	0	Random	
hrc34	Uncomp.			20 % dur.		Freezing
hrc35	Uncomp.				Medium blurring	
hrc36	H.264	1	16	4	Random	Slicing

Table B.6 *Test*3 test plan, video, HD1080

ID	VCOD	KFR (Hz)	VCBR (Mbps)	Nsl	Ppl	PLD	PLC
1	H.264	1	8	1sl/MBR	0	N/A	N/A
2	H.264	1	16	1sl/MBR	4	Random	Slicing
3	Uncompressed			1sl/MBR		N/A	N/A
4	H.264	1	2	1sl/MBR	0	N/A	N/A
5	Uncompressed			1sl/MBR	20 % dur.	N/A	Frozen
6	Uncompressed			1sl/MBR		Medium blurring	N/A
7	H.264	0.125	16	1sl/MBR	0	N/A	N/A
8	H.264	24 (25)	4	1sl/MBR	0	N/A	N/A
9	H.264	1	6	1sl/MBR	0	N/A	N/A
10	H.264	0.5	6	1sl/MBR	0	N/A	N/A
11	MPEG2	1	6	1sl/MBR	0	N/A	N/A
12	H.264	1	16	1sl/MBR	0.25	Bursty ($\mu = 4$)	Slicing
13	H.264	1	16	1sl/MBR	0.5	Bursty ($\mu = 4$)	Slicing
14	H.264	1	16	1sl/MBR	1	Bursty ($\mu = 4$)	Slicing
15	H.264	1	16	1sl/MBR	2	Bursty ($\mu = 4$)	Slicing
16	H.264	1	16	1sl/MBR	4	Bursty ($\mu = 4$)	Slicing
17	H.264	1	16	1sl/MBR	0.5	Bursty ($\mu = 4$)	Frozen
18	H.264	1	16	1sl/MBR	1	Bursty ($\mu = 4$)	Frozen
19	H.264	1	16	1sl/MBR	2	Bursty ($\mu = 4$)	Frozen
20	H.264	1	16	1sl/MBR	0.25	Bursty ($\mu = 16$)	Slicing
21	H.264	1	16	1sl/MBR	0.5	Bursty ($\mu = 16$)	Slicing
22	H.264	1	16	1sl/MBR	1	Bursty ($\mu = 16$)	Slicing
23	H.264	1	16	1sl/MBR	2	Bursty ($\mu = 16$)	Slicing
24	H.264	1	16	1sl/MBR	4	Bursty ($\mu = 16$)	Slicing
25	H.264	1	16	1sl/MBR	0.5	Bursty ($\mu = 16$)	Frozen
26	H.264	1	16	1sl/MBR	1	Bursty ($\mu = 16$)	Frozen
27	H.264	1	16	1sl/MBR	2	Bursty ($\mu = 16$)	Frozen
28	H.264	1	16	1sl/MBR	0.5	Bursty ($\mu = 64$)	Slicing
29	H.264	1	16	1sl/MBR	1	Bursty ($\mu = 64$)	Slicing
30	H.264	1	16	1sl/MBR	2	Bursty ($\mu = 64$)	Slicing
31	H.264	1	16	1sl/MBR	4	Bursty ($\mu = 64$)	Slicing
32	H.264	1	16	1sl/MBR	0.5	Bursty ($\mu = 64$)	Frozen
33	H.264	1	16	1sl/MBR	1	Bursty ($\mu = 64$)	Frozen
34	H.264	1	16	1sl/MBR	2	Bursty ($\mu = 64$)	Frozen
35	H.264	1	4	1sl/MBR	0.5	Bursty ($\mu = 4$)	Slicing
36	H.264	1	4	1sl/MBR	2	Bursty ($\mu = 4$)	Slicing
37	H.264	1	16	3sl/f	0.25	Bursty ($\mu = 4$)	Slicing
38	H.264	1	16	3sl/f	0.5	Bursty ($\mu = 4$)	Slicing
39	H.264	1	16	3sl/f	1	Bursty ($\mu = 4$)	Slicing
40	H.264	1	16	3sl/f	2	Bursty ($\mu = 4$)	Slicing
41	H.264	1	16	1sl/MBR	0.125	Random	Slicing
42	H.264	1	16	1sl/MBR	0.5	Random	Slicing
43	H.264	1	16	1sl/MBR	2	Random	Slicing
44	H.264	1	16	1sl/MBR	0.06	Random	Frozen

(continued)

Table B.6 (continued)

ID	VCOD	KFR (Hz)	VCBR (Mbps)	Nsl	Ppl	PLD	PLC
45	H.264	1	16	1sl/MBR	0.25	Random	Frozen
46	H.264	1	16	1sl/MBR	1	Random	Frozen
47	H.264	1	16	3sl/f	0.25	Random	Slicing
48	H.264	1	16	3sl/f	0.5	Random	Slicing
49	H.264	1	16	3sl/f	1	Random	Slicing
50	H.264	1	16	3sl/f	2	Random	Slicing
51	H.264	1	16	1sl/MBR		Top of I-Frame 7+8 of GOP	Slicing
52	H.264	1	16	1sl/MBR		Top of I-Frame 7+8 of GOP	Frozen
53	H.264	1	16	1sl/MBR		Middle of I-Frame 7+8 of GOP	Slicing
54	H.264	1	16	1sl/MBR		Top of I-Frame 5+10 of GOP	Frozen
55	H.264	1	16	1sl/MBR		Middle of I-Frame 5+10 of GOP	Slicing

$1sl/MBR$: 1 slice per Macroblock-Row; $3sl/f$: 3 slices per frame

Table B.7 *Test*4 test plan, video, HD1080

ID	VCOD	KFR (Hz)	VCBR (Mbps)	Ppl	PLD	PLC
hrc01	H.264	1	8	0		
hrc02	H.264	1	16	4	Random	Slicing
hrc03	Uncomp.					
hrc04	H.264	1	2	0		
hrc05	Uncomp.			20 % dur.		Freezing
hrc06	Uncomp.				Medium blurring 8:0.4:0	
hrc07	H.264	1	16	0		
hrc08	H.264	1	16	0.25	Random	Slicing
hrc09	H.264	1	16	1	Random	Slicing
hrc10	H.264	1	16	4	Random	Slicing
hrc11	H.264	1	4	0		
hrc12	H.264	1	16	0.06	Random	Freezing
hrc13	H.264	1	16	0.25	Random	Freezing
hrc14	H.264	1	16	1	Random	Freezing
hrc15	H.264	1	16	0.25	Random	Slicing
hrc16	H.264	1	16	1	Random	Slicing
hrc17	H.264	1	16	4	Random	Slicing
hrc18	H.264	1	4	0.25	Random	Slicing
hrc19	H.264	1	4	1	Random	Slicing
hrc20	H.264	1	4	4	Random	Slicing
hrc21	H.264	1	16	0	Random	Highprofile
hrc22	H.264	1	8	0	Random	Highprofile
hrc23	H.264	1	4	0	Random	Highprofile

(continued)

Table B.7 (continued)

ID	VCOD	KFR (Hz)	VCBR (Mbps)	Ppl	PLD	PLC
hrc24	H.264	8	16	0		
hrc25	H.264	2	16	0		
hrc26	H.264	8	16	0.25	Random	Slicing
hrc27	H.264	8	16	2	Random	Slicing
hrc28	H.264	8	16	0.06	Random	Freezing
hrc29	H.264	8	16	1	Random	Freezing
hrc30	H.264	2	16	0.25	Random	Slicing
hrc31	H.264	2	16	2	Random	Slicing
hrc32	H.264	2	16	0.06	Random	Freezing
hrc33	H.264	2	16	1	Random	Freezing
hrc34	H.264	8	4	0.25	Random	Slicing
hrc35	H.264	8	4	2	Random	Slicing
hrc36	H.264	2	4	0.25	Random	Slicing
hrc37	H.264	2	4	2	Random	Slicing
hrc38	H.264	1	16	0.06	3-slice-random	Slicing
hrc39	H.264	1	16	0.25	3-slice-random	Slicing
hrc40	H.264	1	16	1	3-slice-random	Slicing
hrc41	H.264	1	4	0.06	3-slice-random	Slicing
hrc42	H.264	1	4	0.25	3-slice-random	Slicing
hrc43	H.264	1	4	1	3-slice-random	Slicing

Table B.8 *Test*5 test plan, video, HD1080

ID	COD	KFR (Hz)	VCBR	Ppl	PLD	PLC
hrc01	H.264	1	16	0	Random	N/A
hrc02	H.264	1	8	0	Random	N/A
hrc03	H.264	1	4	0	Random	N/A
hrc04	H.264	1	16	0.02	Random	Freezing
hrc05	H.264	1	16	0.06	Random	Freezing
hrc06	H.264	1	16	0.125	Random	Freezing
hrc07	H.264	1	16	0.25	Random	Freezing
hrc08	H.264	1	16	0.125	Random	Slicing
hrc09	H.264	1	16	0.25	Random	Slicing
hrc10	H.264	1	16	0.5	Random	Slicing
hrc11	H.264	1	16	1	Random	Slicing
hrc12	H.264	1	16	2	Random	Slicing
hrc13	H.264	1	16	4	Random	Slicing
hrc14	H.264	1	8	0.02	Random	Freezing
hrc15	H.264	1	8	0.06	Random	Freezing
hrc16	H.264	1	8	0.125	Random	Freezing
hrc17	H.264	1	8	0.125	Random	Slicing
hrc18	H.264	1	8	0.5	Random	Slicing
hrc19	H.264	1	8	2	Random	Slicing
hrc20	H.264	1	4	0.02	Random	Freezing
hrc21	H.264	1	4	0.06	Random	Freezing
hrc22	H.264	1	4	0.125	Random	Freezing
hrc23	H.264	1	4	0.25	Random	Freezing
hrc24	H.264	1	4	0.125	Random	Slicing
hrc25	H.264	1	4	0.25	Random	Slicing
hrc26	H.264	1	4	0.5	Random	Slicing
hrc27	H.264	1	4	1	Random	Slicing
hrc28	H.264	1	4	2	Random	Slicing
hrc29	H.264	1	4	4	Random	Slicing
hrc30	Uncomp.					
hrc31	H.264	1	2	0	Random	N/A
hrc32	Uncomp.			20 % dur.		Freezing
hrc33	Uncomp.				Medium blurring	N/A
hrc34	H.264	1	16	4	Random	Slicing

Table B.9 *Ref* test plan, video, SD PAL. *dur.*: duration

ID	VCOD	KFR (Hz)	VCBR	Ppl	PLD	PLC
hrc01	Uncomp.					
hrc02	H.264	1	8			
hrc03	H.264	1	4			
hrc04	H.264	1	2			
hrc05	H.264	1	1			
hrc06	H.264	1	0.5			
hrc07	MPEG-2	1	16			
hrc08	MPEG-2	1	8			
hrc09	MPEG-2	1	4			
hrc10	MPEG-2	1	2			
hrc11	MPEG-2	1	1			
hrc12	Uncomp.			5 % dur.		Freezing
hrc13	Uncomp.			10 % dur.		Freezing
hrc14	Uncomp.			20 % dur.		Freezing
hrc15	Uncomp.			1 out of 2 frames		Freezing
hrc16	Uncomp.			1 out of 5 frames		Freezing
hrc17	Uncomp.				Slight blurring	
hrc18	Uncomp.				Medium blurring	
hrc19	Uncomp.				Strong blurring	
hrc20	H.264	10	16			
hrc21	H.264	25	16			
hrc22	H.264	1	16	1		Slicing
hrc23	H.264	1	16	2		Slicing
hrc24	H.264	1	16	4		Slicing

Table B.10 Test plan used for *Test*1 and *Test*2, video, SD PAL

ID	VCOD	KFR (Hz)	VCBR	Ppl	PLD	PLC
hrc01	H.264	1	4	0	Random	
hrc02	H.264	1	2	0	Random	
hrc03	H.264	1	1	0	Random	
hrc04	H.264	1	4	0.06	Random	Freezing
hrc05	H.264	1	4	0.125	Random	Freezing
hrc06	H.264	1	4	0.25	Random	Freezing
hrc07	H.264	1	4	0.5	Random	Freezing
hrc08	H.264	1	4	0.125	Random	Slicing
hrc09	H.264	1	4	0.25	Random	Slicing
hrc10	H.264	1	4	0.5	Random	Slicing
hrc11	H.264	1	4	1	Random	Slicing
hrc12	H.264	1	4	2	Random	Slicing
hrc13	H.264	1	4	4	Random	Slicing
hrc14	H.264	1	2	0.06	Random	Freezing
hrc15	H.264	1	2	0.125	Random	Freezing
hrc16	H.264	1	2	0.25	Random	Freezing
hrc17	H.264	1	2	0.125	Random	Slicing
hrc18	H.264	1	2	0.5	Random	Slicing
hrc19	H.264	1	2	2	Random	Slicing
hrc20	H.264	1	2	4	Random	Slicing
hrc21	H.264	1	1	0.06	Random	Freezing
hrc22	H.264	1	1	0.125	Random	Freezing
hrc23	H.264	1	1	0.25	Random	Freezing
hrc24	H.264	1	1	0.125	Random	Slicing
hrc25	H.264	1	1	0.25	Random	Slicing
hrc26	H.264	1	1	0.5	Random	Slicing
hrc27	H.264	1	1	1	Random	Slicing
hrc28	H.264	1	1	2	Random	Slicing
hrc29	H.264	1	1	4	Random	Slicing
hrc30	Uncomp.					
hrc31	H.264	1	0.5	0	Random	
hrc32	Uncomp.			20 % dur.		Freezing
hrc33	Uncomp.				Medium blurring	
hrc34	H.264	1	4	4	Random	Slicing

Table B.11 *Test*3 test plan, video, SD PAL

ID	VCOD	KFR (Hz)	VCBR (Mbps)	Nsl	Ppl	PLD	PLC
1	H.264	1	2	1sl/MBR	0	N/A	N/A
2	H.264	1	4	1sl/MBR	4	Random	Slicing
3	Uncompressed			1sl/MBR		N/A	N/A
4	H.264	1	0.5	1sl/MBR	0	N/A	N/A
5	Uncompressed			1sl/MBR	20 % dur.	N/A	Frozen
6	Uncompressed			1sl/MBR		Medium blurring	N/A
7	H.264	0.125	4	1sl/MBR	0	N/A	N/A
8	H.264	24 (25)	1	1sl/MBR	0	N/A	N/A
9	H.264	1	3	1sl/MBR	0	N/A	N/A
10	H.264	0.5	3	1sl/MBR	0	N/A	N/A
11	MPEG2	1	3	1sl/MBR	0	N/A	N/A
12	H.264	1	4	1sl/MBR	0.25	Bursty ($\mu = 4$)	Slicing
13	H.264	1	4	1sl/MBR	0.5	Bursty ($\mu = 4$)	Slicing
14	H.264	1	4	1sl/MBR	1	Bursty ($\mu = 4$)	Slicing
15	H.264	1	4	1sl/MBR	2	Bursty ($\mu = 4$)	Slicing
16	H.264	1	4	1sl/MBR	4	Bursty ($\mu = 4$)	Slicing
17	H.264	1	4	1sl/MBR	0.5	Bursty ($\mu = 4$)	Frozen
18	H.264	1	4	1sl/MBR	1	Bursty ($\mu = 4$)	Frozen
19	H.264	1	4	1sl/MBR	2	Bursty ($\mu = 4$)	Frozen
20	H.264	1	4	1sl/MBR	0.25	Bursty ($\mu = 16$)	Slicing
21	H.264	1	4	1sl/MBR	0.5	Bursty ($\mu = 16$)	Slicing
22	H.264	1	4	1sl/MBR	1	Bursty ($\mu = 16$)	Slicing
23	H.264	1	4	1sl/MBR	2	Bursty ($\mu = 16$)	Slicing
24	H.264	1	4	1sl/MBR	4	Bursty ($\mu = 16$)	Slicing
25	H.264	1	4	1sl/MBR	0.5	Bursty ($\mu = 16$)	Frozen
26	H.264	1	4	1sl/MBR	1	Bursty ($\mu = 16$)	Frozen
27	H.264	1	4	1sl/MBR	2	Bursty ($\mu = 16$)	Frozen
28	H.264	1	4	1sl/MBR	0.5	Bursty ($\mu = 64$)	Slicing
29	H.264	1	4	1sl/MBR	1	Bursty ($\mu = 64$)	Slicing
30	H.264	1	4	1sl/MBR	2	Bursty ($\mu = 64$)	Slicing
31	H.264	1	4	1sl/MBR	4	Bursty ($\mu = 64$)	Slicing
32	H.264	1	4	1sl/MBR	0.5	Bursty ($\mu = 64$)	Frozen
33	H.264	1	4	1sl/MBR	1	Bursty ($\mu = 64$)	Frozen
34	H.264	1	4	1sl/MBR	2	Bursty ($\mu = 64$)	Frozen
35	H.264	1	1	1sl/MBR	0.5	Bursty ($\mu = 4$)	Slicing
36	H.264	1	1	1sl/MBR	2	Bursty ($\mu = 4$)	Slicing
37	H.264	1	4	3sl/f	0.25	Bursty ($\mu = 4$)	Slicing
38	H.264	1	4	3sl/f	0.5	Bursty ($\mu = 4$)	Slicing
39	H.264	1	4	3sl/f	1	Bursty ($\mu = 4$)	Slicing

(continued)

Table B.11 (continued)

ID	VCOD	KFR (Hz)	VCBR (Mbps)	Nsl	Ppl	PLD	PLC
40	H.264	1	4	3sl/f	2	Bursty ($\mu = 4$)	Slicing
41	H.264	1	4	1sl/MBR	0.125	Random	Slicing
42	H.264	1	4	1sl/MBR	0.5	Random	Slicing
43	H.264	1	4	1sl/MBR	2	Random	Slicing
44	H.264	1	4	1sl/MBR	0.06	Random	Frozen
45	H.264	1	4	1sl/MBR	0.25	Random	Frozen
46	H.264	1	4	1sl/MBR	1	Random	Frozen
47	H.264	1	4	3sl/f	0.25	Random	Slicing
48	H.264	1	4	3sl/f	0.5	Random	Slicing
49	H.264	1	4	3sl/f	1	Random	Slicing
50	H.264	1	4	3sl/f	2	Random	Slicing
51	H.264	1	4	1sl/MBR		Top of I-Frame 7+8 of GOP	Slicing
52	H.264	1	4	1sl/MBR		Top of I-Frame 7+8 of GOP	Frozen
53	H.264	1	4	1sl/MBR		Middle of I-Frame 7+8 of GOP	Slicing
54	H.264	1	4	1sl/MBR		Top of I-Frame 5+10 of GOP	Frozen
55	H.264	1	4	1sl/MBR		Middle of I-Frame 5+10 of GOP	Slicing

$1sl/MBR$: 1 slice per Macroblock-row; $3sl/f$: 3 slices per frame

Table B.12 *Test*4 test plan, video, SD PAL

ID	VCOD	KFR (Hz)	VCBR (Mbps)	Ppl	PLD	PLC
hrc01	H.264	1	2	0		
hrc02	H.264	1	4	4	Random	Slicing
hrc03	Uncomp.					
hrc04	H.264	1	0.5	0		
hrc05	Uncomp.			20 % dur.		Freezing
hrc06	Uncomp.				Medium blurring 8:0.4:0	
hrc07	H.264	1	4	0		
hrc08	H.264	1	4	0.25	Random	Slicing
hrc09	H.264	1	4	1	Random	Slicing
hrc10	H.264	1	4	4	Random	Slicing
hrc11	H.264	1	1	0		
hrc12	H.264	1	4	0.06	Random	Freezing
hrc13	H.264	1	4	0.25	Random	Freezing
hrc14	H.264	1	4	1	Random	Freezing
hrc15	H.264	1	4	0.25	Random	Slicing
hrc16	H.264	1	4	1	Random	Slicing
hrc17	H.264	1	4	4	Random	Slicing
hrc18	H.264	1	1	0.25	Random	Slicing
hrc19	H.264	1	1	1	Random	Slicing
hrc20	H.264	1	1	4	Random	Slicing
hrc21	H.264	1	4	0	Random	Highprofile
hrc22	H.264	1	2	0	Random	Highprofile
hrc23	H.264	1	1	0	Random	Highprofile
hrc24	H.264	8	4	0		
hrc25	H.264	2	4	0		
hrc26	H.264	8	4	0.25	Random	Slicing
hrc27	H.264	8	4	2	Random	Slicing
hrc28	H.264	8	4	0.06	Random	Freezing
hrc29	H.264	8	4	1	Random	Freezing
hrc30	H.264	2	4	0.25	Random	Slicing
hrc31	H.264	2	4	2	Random	Slicing
hrc32	H.264	2	4	0.06	Random	Freezing
hrc33	H.264	2	4	1	Random	Freezing
hrc34	H.264	8	1	0.25	Random	Slicing
hrc35	H.264	8	1	2	Random	Slicing
hrc36	H.264	2	1	0.25	Random	Slicing
hrc37	H.264	2	1	2	Random	Slicing
hrc38	H.264	1	4	0.06	3-slice-random	Slicing
hrc39	H.264	1	4	0.25	3-slice-random	Slicing
hrc40	H.264	1	4	1	3-slice-random	Slicing
hrc41	H.264	1	1	0.06	3-slice-random	Slicing
hrc42	H.264	1	1	0.25	3-slice-random	Slicing
hrc43	H.264	1	1	1	3-slice-random	Slicing

Table B.13 *Test*5 test plan, video, SD PAL

ID	COD	KFR (Hz)	VCBR	Ppl	PLD	PLC
hrc01	H.264	1	4	0	Random	N/A
hrc02	H.264	1	2	0	Random	N/A
hrc03	H.264	1	1	0	Random	N/A
hrc04	H.264	1	4	0.06	Random	Freezing
hrc05	H.264	1	4	0.125	Random	Freezing
hrc06	H.264	1	4	0.25	Random	Freezing
hrc07	H.264	1	4	0.5	Random	Freezing
hrc08	H.264	1	4	0.125	Random	Slicing
hrc09	H.264	1	4	0.25	Random	Slicing
hrc10	H.264	1	4	0.5	Random	Slicing
hrc11	H.264	1	4	1	Random	Slicing
hrc12	H.264	1	4	2	Random	Slicing
hrc13	H.264	1	4	4	Random	Slicing
hrc14	H.264	1	2	0.06	Random	Freezing
hrc15	H.264	1	2	0.125	Random	Freezing
hrc16	H.264	1	2	0.25	Random	Freezing
hrc17	H.264	1	2	0.125	Random	Slicing
hrc18	H.264	1	2	0.5	Random	Slicing
hrc19	H.264	1	2	2	Random	Slicing
hrc20	H.264	1	1	0.06	Random	Freezing
hrc21	H.264	1	1	0.125	Random	Freezing
hrc22	H.264	1	1	0.25	Random	Freezing
hrc23	H.264	1	1	0.5	Random	Freezing
hrc24	H.264	1	1	0.125	Random	Slicing
hrc25	H.264	1	1	0.25	Random	Slicing
hrc26	H.264	1	1	0.5	Random	Slicing
hrc27	H.264	1	1	1	Random	Slicing
hrc28	H.264	1	1	2	Random	Slicing
hrc29	H.264	1	1	4	Random	Slicing
hrc30	Uncomp.					
hrc31	H.264	1	0.5	0	Random	N/A
hrc32	Uncomp.			20 % dur.		Freezing
hrc33	Uncomp.				Medium blurring	N/A
hrc34	H.264	1	4	4	Random	Slicing

Table B.14 $ph2_{AV}$ test plan, audiovisual, HD1080

ID	ACBR (kbps)	VCBR (Mbps)	Pfl	Ppl	PLC
hrc01	Uncomp.	Uncomp.	0	0	N/A
hrc02	Uncomp.	16	0	4	Slicing
hrc03	Uncomp.		0	20 % dur.	Freezing
hrc04	Uncomp.	2	0	0	N/A
hrc05	Uncomp.	8	0	0	N/A
hrc06	BW7	Uncomp.	0	0	N/A
hrc07	192	16	8	4	Slicing
hrc08	48 (AAC-LC)	Uncomp.	0	0	N/A
hrc09	192	Uncomp.	8	0	N/A
hrc10	96	2	0	0	N/A
hrc11	96	4	0	0	N/A
hrc12	96	16	0	0	N/A
hrc13	128	2	0	0	N/A
hrc14	128	4	0	0	N/A
hrc15	128	16	0	0	N/A
hrc16	192	2	0	0	N/A
hrc17	192	4	0	0	N/A
hrc18	192	16	0	0	N/A
hrc19	192	16	0	0.125	Slicing
hrc20	192	16	0	0.5	Slicing
hrc21	192	4	0	4	Slicing
hrc22	192	16	0	0.06	Frozen
hrc23	192	16	0	0.25	Frozen
hrc24	192	16	0	1	Frozen
hrc25	192	16	1	0	N/A
hrc26	192	16	1	0.125	Slicing
hrc27	192	16	1	0.5	Slicing
hrc28	192	4	1	4	Slicing
hrc29	192	16	1	0.06	Frozen
hrc30	192	16	1	0.25	Frozen
hrc31	192	16	1	1	Frozen
hrc32	192	16	4	0	N/A
hrc33	192	16	4	0.125	Slicing
hrc34	192	4	4	4	Slicing
hrc35	192	16	4	0.06	Frozen
hrc36	192	16	4	1	Frozen
hrc37	192	16	8	0	N/A
hrc38	192	16	8	0.125	Slicing
hrc39	192	16	8	0.5	Slicing
hrc40	192	4	8	4	Slicing

(continued)

Table B.14 (continued)

ID	ACBR (kbps)	VCBR (Mbps)	Pfl	Ppl	PLC
hrc41	192	16	8	0.06	Frozen
hrc42	192	16	8	0.25	Frozen
hrc43	192	16	8	1	Frozen
hrc44	192	2	1	0	N/A
hrc45	192	4	1	0	N/A
hrc46	96	16	0	0.125	Slicing
hrc47	128	16	0	0.125	Slicing
hrc48	96	16	0	0.06	Frozen
hrc49	128	16	0	0.06	Frozen

Random audio frame loss and video packet-loss. Audio codec: MP2; Video codec: H.264; KFR: 1 Hz

Table B.15 $ph3_{AV}$ test plan, audiovisual, HD1080

ID	ACBR (kbps)	VCBR (Mbps)	Pfl	Ppl	PLC
hrc01	Uncomp.	Uncomp.	0	0	N/A
hrc02	Uncomp.	16	0	4	Slicing
hrc03	Uncomp.	Uncomp.	0	20 % dur.	N/A
hrc04	Uncomp.	2	0	0	N/A
hrc05	Uncomp.	8	0	0	N/A
hrc06	BW7	Uncomp.	0	0	N/A
hrc07	192	16	8	4	Slicing
hrc08	48 (AAC-LC)	Uncomp.	0	0	N/A
hrc09	192	Uncomp.	8	0	N/A
hrc10	48	2	0	0	N/A
hrc11	48	4	0	0	N/A
hrc12	48	16	0	0	N/A
hrc13	96	2	0	0	N/A
hrc14	96	4	0	0	N/A
hrc15	96	16	0	0	N/A
hrc16	192	2	0	0	N/A
hrc17	192	4	0	0	N/A
hrc18	192	16	0	0	N/A
hrc19	192	16	0	0.125	Slicing
hrc20	192	16	0	0.5	Slicing
hrc21	192	16	0	4	Slicing
hrc22	192	16	0	0.02	Freezing
hrc23	192	16	0	0.06	Freezing
hrc24	192	16	0	0.25	Freezing
hrc25	192	16	1	0	N/A
hrc26	192	16	1	0.125	Slicing
hrc27	192	16	1	0.5	Slicing
hrc28	192	16	1	4	Slicing
hrc29	192	16	1	0.02	Freezing
hrc30	192	16	1	0.06	Freezing
hrc31	192	16	1	0.25	Freezing
hrc32	192	16	4	0	N/A
hrc33	192	16	4	0.125	Slicing
hrc34	192	16	4	4	Slicing
hrc35	192	16	4	0.02	Freezing
hrc36	192	16	4	0.25	Freezing
hrc37	192	16	8	0	N/A
hrc38	192	16	8	0.125	Slicing
hrc39	192	16	8	0.5	Slicing

(continued)

Table B.15 (continued)

ID	ACBR (kbps)	VCBR (Mbps)	Pfl	Ppl	PLC
hrc40	192	16	8	4	Slicing
hrc41	192	16	8	0.02	Freezing
hrc42	192	16	8	0.06	Freezing
hrc43	192	16	8	0.25	Freezing
hrc44	192	2	1	0	N/A
hrc45	192	4	1	0	N/A
hrc46	48	16	0	0.125	Slicing
hrc47	96	16	0	0.125	Slicing
hrc48	48	16	0	0.02	Freezing
hrc49	96	16	0	0.02	Freezing

Random audio frame loss and video packet-loss. Audio codec: MP2; Video codec: H.264; KFR: 1 Hz

Table B.16 $ph2_{AV}$ test plan, audiovisual, SD PAL

ID	ACBR (kbps)	VCBR (Mbps)	Pfl	Ppl	PLC
hrc01	Uncomp.	Uncomp.	0	0	N/A
hrc02	Uncomp.	4	0	4	Slicing
hrc03	Uncomp.		0	20 % dur.	Freezing
hrc04	Uncomp.	0.5	0	0	N/A
hrc05	Uncomp.	2	0	0	N/A
hrc06	BW7	Uncomp.	0	0	N/A
hrc07	192	4	8	4	Slicing
hrc08	48 (AAC-LC)	Uncomp.	0	0	N/A
hrc09	192	Uncomp.	8	0	N/A
hrc10	96	0.5	0	0	N/A
hrc11	96	1	0	0	N/A
hrc12	96	4	0	0	N/A
hrc13	128	0.5	0	0	N/A
hrc14	128	1	0	0	N/A
hrc15	128	4	0	0	N/A
hrc16	192	0.5	0	0	N/A
hrc17	192	1	0	0	N/A
hrc18	192	4	0	0	N/A
hrc19	192	4	0	0.125	Slicing
hrc20	192	4	0	0.5	Slicing
hrc21	192	1	0	4	Slicing
hrc22	192	4	0	0.06	Frozen
hrc23	192	4	0	0.25	Frozen
hrc24	192	4	0	1	Frozen
hrc25	192	4	1	0	N/A
hrc26	192	4	1	0.125	Slicing
hrc27	192	4	1	0.5	Slicing
hrc28	192	1	1	4	Slicing
hrc29	192	4	1	0.06	Frozen
hrc30	192	4	1	0.25	Frozen
hrc31	192	4	1	1	Frozen
hrc32	192	4	4	0	N/A
hrc33	192	4	4	0.125	Slicing
hrc34	192	1	4	4	Slicing
hrc35	192	4	4	0.06	Frozen
hrc36	192	4	4	1	Frozen
hrc37	192	4	8	0	N/A
hrc38	192	4	8	0.125	Slicing
hrc39	192	4	8	0.5	Slicing

(continued)

Table B.16 (continued)

ID	ACBR (kbps)	VCBR (Mbps)	Pfl	Ppl	PLC
hrc40	192	1	8	4	Slicing
hrc41	192	4	8	0.06	Frozen
hrc42	192	4	8	0.25	Frozen
hrc43	192	4	8	1	Frozen
hrc44	192	0.5	1	0	N/A
hrc45	192	1	1	0	N/A
hrc46	96	4	0	0.125	Slicing
hrc47	128	4	0	0.125	Slicing
hrc48	96	4	0	0.06	Frozen
hrc49	128	4	0	0.06	Frozen

Random audio frame loss and video packet-loss. Audio codec: MP2; Video codec: H.264; KFR: 1 Hz

Table B.17 $ph3_{AV}$ test plan, audiovisual, SD PAL

ID	ACBR (kbps)	VCBR (Mbps)	Pfl	Ppl	PLC
hrc01	Uncomp.	Uncomp.	0	0	N/A
hrc02	Uncomp.	4	0	4	Slicing
hrc03	Uncomp.	Uncomp.	0	20 % dur.	N/A
hrc04	Uncomp.	0.5	0	0	N/A
hrc05	Uncomp.	2	0	0	N/A
hrc06	BW7	Uncomp.	0	0	N/A
hrc07	192	4	8	4	Slicing
hrc08	48 (AAC-LC)	Uncomp.	0	0	N/A
hrc09	192	Uncomp.	8	0	N/A
hrc10	48	0.5	0	0	N/A
hrc11	48	1	0	0	N/A
hrc12	48	4	0	0	N/A
hrc13	96	0.5	0	0	N/A
hrc14	96	1	0	0	N/A
hrc15	96	4	0	0	N/A
hrc16	192	0.5	0	0	N/A
hrc17	192	1	0	0	N/A
hrc18	192	4	0	0	N/A
hrc19	192	4	0	0.125	Slicing
hrc20	192	4	0	0.5	Slicing
hrc21	192	4	0	4	Slicing
hrc22	192	4	0	0.06	Freezing
hrc23	192	4	0	0.25	Freezing
hrc24	192	4	0	1	Freezing
hrc25	192	4	1	0	N/A
hrc26	192	4	1	0.125	Slicing
hrc27	192	4	1	0.5	Slicing
hrc28	192	4	1	4	Slicing
hrc29	192	4	1	0.06	Freezing
hrc30	192	4	1	0.25	Freezing
hrc31	192	4	1	1	Freezing
hrc32	192	4	4	0	N/A
hrc33	192	4	4	0.125	Slicing
hrc34	192	4	4	4	Slicing
hrc35	192	4	4	0.06	Freezing
hrc36	192	4	4	1	Freezing
hrc37	192	4	8	0	N/A
hrc38	192	4	8	0.125	Slicing
hrc39	192	4	8	0.5	Slicing

(continued)

Table B.17 (continued)

ID	ACBR (kbps)	VCBR (Mbps)	Pfl	Ppl	PLC
hrc40	192	4	8	4	Slicing
hrc41	192	4	8	0.06	Freezing
hrc42	192	4	8	0.25	Freezing
hrc43	192	4	8	1	Freezing
hrc44	192	0.5	1	0	N/A
hrc45	192	1	1	0	N/A
hrc46	48	4	0	0.125	Slicing
hrc47	96	4	0	0.125	Slicing
hrc48	48	4	0	0.02	Freezing
hrc49	96	4	0	0.02	Freezing

Random audio frame loss and video packet-loss. Audio codec: MP2; Video codec: H.264; KFR: 1 *Hz*

Table B.18 *tr*16 test plan, audio

ID	VCOD	ACOD	ACBR	Ppl	PLD
hrc01	H.264	AAC-LC	576	0	0
hrc02	H.264	AAC-LC	96	0	0
hrc03	H.264	MP2	64	0	0
hrc04	H.264	AAC-LC	256	1	Bursty
hrc05	H.264	MP2	384	5	Bursty
hrc06	H.264	AAC-LC	256	0	0
hrc07	H.264	AAC-LC	48	0	0
hrc08	H.264	MP2	384	0	0
hrc09	H.264	MP2	96	0	0
hrc10	H.264	HE-AAC	96	0	0
hrc11	H.264	HE-AAC	48	0	0
hrc12	H.264	HE-AAC	16	0	0
hrc13	H.264	AC3	256	0	0
hrc14	H.264	AC3	128	0	0
hrc15	H.264	AC3	96	0	0
hrc16	H.264	AAC-LC	256	0.5	Bursty
hrc17	H.264	AAC-LC	256	5	Bursty
hrc18	H.264	AAC-LC	48	0.5	Bursty
hrc19	H.264	AAC-LC	48	3	Bursty
hrc20	H.264	MP2	384	0.5	Bursty
hrc21	H.264	MP2	384	1	Bursty
hrc22	H.264	MP2	64	0.5	Bursty
hrc23	H.264	MP2	64	3	Bursty
hrc24	H.264	AC3	256	0.5	Bursty
hrc25	H.264	AC3	256	1	Bursty
hrc26	H.264	AC3	256	1	Random
hrc27	H.264	AC3	256	3	Bursty
hrc28	H.264	AC3	96	0.5	Bursty
hrc29	H.264	AC3	96	1	Bursty
hrc30	H.264	AC3	96	3	Bursty
hrc31	H.264	HE-AAC	96	1	Bursty
hrc32	H.264	HE-AAC	96	3	Bursty
hrc33	H.264	HE-AAC	16	1	Bursty
hrc34	H.264	HE-AAC	16	3	Bursty

Table B.19 *tr*10 test plan, video, HD720

ID	COD	CBR (Mbps)	GOP	Ppl	PLC	PLD
hrc01	H.264	30	M3N50	0	0	0
hrc02	H.264	5	M3N50	0	0	0
hrc03	H.264	15	M3N50	0.125	Slicing	Bursty
hrc04	H.264	5	M3N50	0.5	Slicing	Bursty
hrc05	H.264	15	M3N50	0.5	Freezing	Bursty
hrc06	H.264	15	M3N50	0	0	0
hrc07	H.264	2	M3N50	0	0	0
hrc08	H.264	1	M3N50	0	0	0
hrc09	H.264	0.5	M3N50	0	0	0
hrc10	H.264	15	M3N25	0	0	0
hrc11	H.264	2	M3N25	0	0	0
hrc12	H.264	0.5	M3N25	0	0	0
hrc13	H.264	15	M4N50	0	0	0
hrc14	H.264	2	M4N50	0	0	0
hrc15	H.264	0.5	M4N50	0	0	0
hrc16	H.264	15	M3N50	0.125	Slicing	Random
hrc17	H.264	15	M3N25	0.125	Slicing	Bursty
hrc18	H.264	15	M3N50	0.25	Slicing	Bursty
hrc19	H.264	15	M3N50	0.25	Slicing	Random
hrc20	H.264	15	M3N25	0.25	Slicing	Bursty
hrc21	H.264	15	M3N25	0.25	Slicing	Random
hrc22	H.264	15	M3N50	0.5	Slicing	Bursty
hrc23	H.264	15	M3N50	1	Slicing	Bursty
hrc24	H.264	5	M3N50	0.25	Slicing	Bursty
hrc25	H.264	5	M3N50	1	Slicing	Bursty
hrc26	H.264	15	M3N50	0.06	Freezing	Bursty
hrc27	H.264	15	M3N50	0.125	Freezing	Bursty
hrc28	H.264	15	M3N50	0.25	Freezing	Bursty
hrc29	H.264	5	M3N50	0.125	Freezing	Bursty
hrc30	H.264	5	M3N50	0.25	Freezing	Bursty

Table B.20 *tr*13 test plan, audiovisual, HD720

ID	VCOD	ACOD	VCBR	ACBR	Ppl	PLC	PLD
hrc01	H.264	AAC-LC	30	576	0	N/A	0
hrc02	H.264	MP2	2	96	0	N/A	0
hrc03	H.264	AAC-LC	15	48	0.25	Slicing	Bursty
hrc04	H.264	MP2	15	384	0.5	Slicing	Bursty
hrc05	H.264	MP2	15	384	0.5	Freezing	Bursty
hrc06	H.264	MP2	15	384	0	N/A	0
hrc07	H.264	MP2	15	64	0	N/A	0
hrc08	H.264	MP2	1	384	0	N/A	0
hrc09	H.264	MP2	1	64	0	N/A	0
hrc10	H.264	AAC-LC	15	256	0	N/A	0
hrc11	H.264	AAC-LC	15	48	0	N/A	0
hrc12	H.264	AAC-LC	1	256	0	N/A	0
hrc13	H.264	AAC-LC	1	48	0	N/A	0
hrc14	H.264	MP2	15	384	0.25	Slicing	Bursty
hrc15	H.264	MP2	15	384	1	Slicing	Bursty
hrc16	H.264	MP2	15	64	0.25	Slicing	Bursty
hrc17	H.264	MP2	15	64	0.5	Slicing	Bursty
hrc18	H.264	MP2	2	96	0.25	Slicing	Bursty
hrc19	H.264	MP2	2	96	0.5	Slicing	Bursty
hrc20	H.264	MP2	1	384	0.25	Slicing	Bursty
hrc21	H.264	MP2	1	384	0.5	Slicing	Bursty
hrc22	H.264	AAC-LC	15	256	0.25	Slicing	Bursty
hrc23	H.264	AAC-LC	15	256	0.5	Slicing	Bursty
hrc24	H.264	AAC-LC	15	48	0.5	Slicing	Bursty
hrc25	H.264	MP2	15	384	0.125	Freezing	Bursty
hrc26	H.264	MP2	15	384	0.25	Freezing	Bursty
hrc27	H.264	MP2	15	64	0.125	Freezing	Bursty
hrc28	H.264	MP2	15	64	0.25	Freezing	Bursty
hrc29	H.264	MP2	2	96	0.125	Freezing	Bursty
hrc30	H.264	MP2	2	96	0.25	Freezing	Bursty

References

3GPP TS.26.402: General audio codec audio processing functions; enhanced aacPlus general audio codec; additional decoder tools

Advanced Television Systems Committee. Digital audio compression standard (AC-3, E-AC-3). In: Document A/52:2010

Aldridge RP, Hands DS, Pearson DE, Lodge NK (1998) Continuous quality assessment of digitally-coded television pictures. In: IEE Proceedings of vision, image and signal processing

Allnatt J (1983) Transmitted-picture assessment. Wiley, Chichester

Amelang M, Bartussek D, Stemmler G, Hagemann D (2006) Differentielle psychologie und Persönlichkeitsforschung. W. Kohlhammer Verlag, Stuttgart

Argyropoulos S, List P, Raake A, Feiten B, Garcia MN (2013) Scene change detection for perceptual quality evaluation in video sequences. In: Proceedings of international conference on image processing (ICIP)

Bech S, Zacharov N (2006) Perceptual audio evaluation-theory, method and application. Wiley, Chichester

Beerends JG, Caluwe FE (1999) The influence of video quality on perceived audio quality and vice versa. J Audio Eng Soc 47(5):355–362

Belmudez B, Möller S, Lewcio B, Raake A, Mehmood A (2009) Audio and video channel impact on perceived audio-visual quality in different interactive contexts. In: Proceedings of IEEE international workshop on multimedia, signal processing (MMSP)

Blauert J, Jekosch U (1997) Sound-quality evaluation—a multi-layered problem. Acta Acustica uniter with Acustica 83:747–753

Bodden M, Jekosch U (1996) Entwicklung und durchfuhrung von tests mit versuchspersonen zur verifizierung von modellen zur berechnung der sprach ubertragungsqualitat. In: Final report to a project funded by Deutsche Telekom AG (unpublished), Institut fur Kommunikationsakustik, Ruhr Universitat, Bochum

Boev A, Hollosi D, Gotchev A, Egiazarian K (2009) Classification and simulation of stereoscopic artifacts in mobile 3DTV content. In: Proceedings of SPIE electronic, imaging

Bolot JC, Fosse-Parisis S, Towsley D (1999) Adaptive FEC-based error control for internet telephony. In: Proceedings of IEEE international conference on computer communications (INFOCOM)

Bosi M, Brandenburg K, Quackenbush S, Fielder L, Akagiri K, Fuchs H, Dietz M, Herre J, Davidson G, Oikawa (1996) MPEG 2 advanced audio coding. In: Proceedings of 101st Audio Engineering Society (AES) convention

T-Labs Series in Telecommunication Services, DOI: 10.1007/978-3-319-04855-0,

Boulos F, Parrein B, Le Callet P, Hands DS (2009) Perceptual effects of packet loss on H.264/AVC encoded videos. In: Proceedings of international workshop on video processing and quality metrics for consumer electronics (VPQM)
Brandenburg K (1999) MP3 and AAC explained. In: Proceedings of 17th Audio Engineering Society (AES) conference high quality audio coding
Brandenburg K, Stoll G, Dehéry YF, Johnston JD, Kerkhof LVD, Schroeder EF (1992) The ISO/MPEG-audio codec: a generic standard for coding of high quality digital audio. In: Proceedings of 92th Audio Engineering Society (AES) convention
Broom S (2006) VoIP: quality assessment: taking account of the edge-device. IEEE Trans Audio Speech Lang Process 14(6): 1977–1983
Brotherton MD, Huynh-Thu Q, Hands DS, Brunnstrom K (2005) Subjective multimedia quality assessment. IEICE Trans
Carroll J (1972) Individual differences and multidimensional scaling. In: Shepard RN, Romney AK, Nerlove SB (eds) Multidimensional scaling—theory and application in the behavioral sciences, vol I. Seminar Press, New York
Chateau N (1998) Relations between audio, video and audio-visual quality. In: Contr COM 12–61 to ITU-T Study, Group 12
Chikkerur S, Sundaram V, Reisslein M, Karam LJ (2011) Objective video quality assessment methods: a classification, review, and performance comparison. IEEE Trans Broadcast 57(2):165–182
Choisel S, Wickelmaier F (2007) Evaluation of multichannel reproduced sound: scaling auditory attributes underlying listener preference. J Acoust Soc Am 121:388–400
Clark A (2001) Modeling the effects of burst packet loss and recency on subjective voice quality. In: Internet Telephony Workshop (IPtel)
Clark A (2003) Description of VQMON algorithm. In: ITU-T del. cont. COM12-D105
Coolican H (2009) Research methods and statistics in psychology, 5th edn. Hodder Education, London
Côté N (2010) Integral and diagnostic intrusive prediction of speech quality. Doctoral dissertation
Côté N, Durin V (2008) Effect of degradations distribution in a corpus test on auditory ratings. In: Proceedings of acoustics 08
Côté N, Koehl V, Gautier-Turbin V, Raake A, Möller S (2009) Reference units for the comparison of speech quality test results. In: Proceedings of 126th Audio Engineering Society (AES) convention
Cox EP (1980) The optimal number of response alternatives for a scale: a review. J Mark Res 17:407–422
Davis M (1993) The AC-3 multichannel coder. In: Proceedings of 95th Audio Engineering Society (AES) convention
DIN EN ISO 9000 (2000) Standards related to quality management, 2000–2012
DIN EN ISO 9000: 2005 Supplement (2008) Concepts for quality management—part 11: concepts for quality control and statistics, 2008-05
DIN EN ISO 9000: 2005 Supplement (2008) Concepts for quality management—part 12. Concepts in the field of quality and statictics; concepts relating to characteristics, 2008-05
DSL Forum (2006) Tripleplay services quality of experience (QoE) requirements. In: Architecture & Transport Working Group, technical report TR-126, DSL, Forum, Dec 2006
EBU Tech. 3276 (1998) Listening conditions for the assessment of sound programme material: monophonic and two channel stereophonic
Egi N, Hayashi T, Takahashi A (2010) Parametric packet-layer model for evaluation audio quality in multimedia streaming services. IEICE Trans Commun E93.B:1359–1366
Elliott EO (1960) Estimates of error rates for codes on burst-noise channels. Bell Syst Tech J 42:1977–1997
Engeldrum PG (2004) Psychometric scaling: a toolkit for imaging systems development. J Image Sci Technol 48
Engeldrum PG (2000) Psychometric scaling: a toolkit for imaging systems development. Imcotek Press, Winchester

Erne M (2001) Perceptual audio coders: what to listen for. In: Proceediongs of 111th Audio Engineering Society (AES) convention
Eurescom project p. 807 deliverable 1 (1998) In Jupiter II—usability, performability and interoperability trials in Europe. European Institute for Research and Strategic Studies in Telecommunications
Farias MCQ, Mitra SK (2005) No-reference video quality metric based on artifact measurements. In: Proceedings of international conference on image processing (ICIP)
Farias MCQ (2004) No-reference and reduced reference video quality metrics: new contributions. Doctoral dissertation
Feiten B, Raake A, Garcia MN, Wüstenhagen U, Kroll J (2009) Subjective quality evaluation of audio streaming applications on absolute and paired rating scales. In: Proceedings of 126th Audio Engineering Society (AES) convention
Frossard P, Verscheure O (2001) Joint source/FEC rate selection for quality-optimal MPEG-2 video delivery. IEEE Trans Image Process 10(12):1815–1825
Garcia MN, List P, Argyropoulos S, Lindegren D, Pettersson M, Feiten B, Gustafsson J, Raake A (2013) Parametric model for audiovisual quality assessment in IPTV: ITU-T Rec. P.1201.2. In: Proceedings of the 15th IEEE international workshop on multimedia, signal processing (MMSP)
Garcia MN, Raake A (2009) Impairment-factor based audio-visual quality model for IPTV. In: Proceedings of QOMEX
Garcia MN, Raake A (2010) Normalization of subjective video test results using a reference test and anchor conditions for efficient model development. In: Proceedings of 2nd international workshop on quality of multimedia experience (QoMEX)
Garcia MN, Raake A (2010) Quality impact of video format and scaling in the context of IPTV. In: Proceedings of third international workshop on perceptual quality of systems (PQS)
Garcia MN, Raake A (2011) Frame-layer packet-based parametric video quality model for encrypted video in iptv services. In: Proceedings of international workshop on quality of multimedia experience (QoMEX)
Garcia MN, Raake A (2010) Parametric packet-layer video quality model for IPTV. In: Proceedings of tenth international conference on information science, signal processing and their applications (ISSPA)
Garcia MN, Raake A, Feiten B (2013) Parametric audio quality model for IPTV services—ITU-T P.1201.2 audio. In: Proceedings of the fith international workshop on quality of multimedia experience (QoMEX)
Garcia MN, Schleicher R, Raake A (2010) Towards a content-based parametric video quality model for IPTV. In: Proceedings of 5th international workshop on video processing and quality metrics for consumer electronics (VPQM)
Garcia M-N, Schleicher R, Raake A (2011) Impairment-factor-based audiovisual quality model for IPTV: influence of video resolution, degradation type, and content type. EURASIP J Image Video Process 2011:1–14
Geerts D, De Moor K, Ketyko I, Jacobs A, den Bergh JV, Joseph W, Martens L, De Marez L (2010) Linking an integrated framework for measuring QoE. In: Proceedings of international workshop on quality of multimedia experience (QoMEX)
Gilbert EN (1960) Capacity of a burst-noise channel. Bell Syst Tech J 39:1253–1265
Gleiss N (1987) Improvements of rating methods. In: Document SQ-41.87 to ITU-T speech quality experts group (unpublished), International Telecommunication Union, CH, Geneva
Goldstein EB (2002) Sensation and perception. Wadsworth Publishing, Belmont
Graubner M, Mogre P, Lorenzen T (2010) QoE assessment for audio contribution over IP (ACIP). In: Proceedings of 38th Audio Engineering Society (AES) international conference on sound quality, evaluation
Greengrass J, Evans J, Begen AC (2009) Not all packets are equal: part II. The impact of network packet loss on video quality. IEEE Internet Comput 13(2):74–82
Gros L, Chateau N (2001) Instantaneous and overall judgements for time-varying speech quality: assessments and relationships. Acta Acustica United Acustica 87:367–377

Gros L, Chateau M, Busson S (2004) Effects of context on the subjective assessment of time-varying speech quality: listening/conversation, laboratory/real environment. Acta Acust United Acust 90:1037–1051

Guastavino C, Katz BFG (2004) Perceptual evaluation of multidimensional spatial audio reproduction. J Acoust Soc Am 116:1105–1115

Hall JL (2001) Application of multidimensional scaling to subjective evaluation of coded speech. In: Proceedings of IEEE workshop on speech coding

Hands DS, Avons SE (2001) Recency and duration neglect in subjective assessment of television picture quality. Appl Cogn Sci 15:639–657

Hands DS (2004) A basic multimedia quality model. IEEE Trans Multimedia 6(6):806–816

Hastie T, Tibshirani R, Friedman J (2009) The elements of statistical learning: data mining, inference, and prediction, 2nd edn., StatisticsSpringer, New Work

Herre J (1999) Temporal noise shaping, quantization and coding methods in perceptual audio coding: a tutorial introduction. In: Proceedings of 17th Audio Engineering Society (AES) international conference

Heynderickx I, Muijs R (2005) How perceptual knowledge might affect video processing. In: Proceedings of Workshop on the video picture quality metrics

Hollier MP, Rimell AN, Hands DS, Voelcker RM (1999) Multi-modal perception. BT Technol J 17(1):35–46

http://ffmpeg.mplayerhq.hu

http://sirannon.atlantis.ugent.be

http://www.fourcc.org/yuv.php

http://www.mplayerhq.hu

http://www.nero.com/eng/downloads-nerodigital-nero-aac-codec.php

http://www.videolan.org/developers/x264.html

Huszák Á, Imre S (2010) Analysis GOP structure and packet loss effects on error propagation in MPEG-4 video streams. In: Proceedings of the 4th international symposium on communications, control and signal processing (ISCCP)

Huynh-Thu Q, Garcia MN, Speranza F, Corriveau PJ, Raake A (2011) Study of rating scales for subjective quality assessment of high-definition video. IEEE Trans Broadcast 57(1):1–14

Huynh-Thu Q, Ghanbari M (2005) A comparison of subjective video quality assessment methods for low-bit rate and low-resolution video. In: Proceedings of IASTED international conference on signal and image processing, vol 479, pp 70–76

Huynh-Thu Q, Ghanbari M, Hands DS, Brotherton M (2006) Subjective video quality evaluation for multimedia applications. In: Proceedings of SPIE/IS&T human vision and electronic imaging (HVEI)

Ibarrola E, Xiao J, Liberal F, Ferro A (2011) Internet QoS regulation in future networks: a user-centric approach. IEEE Commun Mag

IETF RFC 2250, RTP Payload Format for MPEG1/MPEG2 Video (1998)

IETF RFC 3550, RTP: A transport protocol for real-time applications

ISO/IEC 11172–3 (1993) Coding of moving pictures and associated audio for digital storage media at up to about 1.5Mbit/s—part 3: audio

ISO/IEC 13818–2 (2000) Generic coding of moving pictures and associated audio: video

ISO/IEC 13818–3 (1995) Generic coding of moving pictures and associated audio: audio

ISO/IEC 13818–7 (2006) Generic coding of moving pictures and associated audio: advanced audio coding

ISO/IEC 14496–10 (2010) Information technology—coding of audio-visual objects—part 10: Advanced video coding

ISO/IEC 14496–3 (2006) Information technology—coding of audio-visual objects—part 3: audio

ITU-R Recommendation BS.1116-1 (1997) Methods for the subjective assessment of small impairments in audio systems including multichannel sound systems

ITU-R Recommendation BS.1534-1 (2003) Method for the subjective assessment of intermediate quality levels of coding systems (MUSHRA)

ITU-R Recommendation BT.1788 (2007) Methodology for the subjective assessment of video quality in multimedia applications
ITU-R Recommendation BT.500-12 (2009) Methodology for the subjective assessment of the quality of television pictures
ITU-R Recommendation BT.601-7 (2011) Studio encoding parameters of digital television for standard 4:3 and wide-screen 16:9 aspect ratios
ITU-R SG12 COM 12–37 (Bellcore, USA) (1994) Extension of combined audio/video quality model
ITU-T Rec. P.10 Amendment 2 (2008) New definitions for inclusion in Recommendation ITU-T P.10/G.100
ITU-T Rec. P.930 (1996) Principles of a reference impairment system for video
Itu-t recommendation bt.710-4: subjective assessment methods for image quality in high-definition television
ITU-T Recommendation G.107 (2005) The E-model, a computational model for use in transmission planning
ITU-T Recommendation P.1201 (2012) Parametric non-intrusive assessment of audiovisual media streaming quality
ITU-T Recommendation P.1201.1 (2012) Parametric non-intrusive assessment of audiovisual media streaming quality—lower resolution application area
ITU-T Recommendation P.1201.2 (2012) Parametric non-intrusive assessment of audiovisual media streaming quality—higher resolution application area
ITU-T Recommendation P.1201/2 Supplement (2013) Processing chain. submitted to ITU-T, expected 2013
ITU-T Recommendation P.1201/2 Supplement (2013) Test Plan. submitted to ITU-T, expected 2013
ITU-T Recommendation P.1202 (2012) Parametric non-intrusive bitstream assessment of video media streaming quality
ITU-T Recommendation P.1202.1 (2012) Parametric non-intrusive bitstream assessment of video media streaming quality—lower resolution application area
ITU-T Recommendation P.1202.2 (2012) Parametric non-intrusive bitstream assessment of video media streaming quality—higher resolution application area
ITU-T Recommendation P.1401 (2012) Methods, metrics and procedures for statistical evaluation, qualification and comparison of objective quality prediction models
ITU-T Recommendation J.144 (2001) Objective perceptual video quality measurement techniques for digital cable television in the presence of a full reference
ITU-T Recommendation J.148 (1993) Tolerances for transmission time differences between the vision and sound components of a television signal
ITU-T Recommendation J.148 (2003) Requirements for an objective perceptual multimedia quality model
ITU-T Recommendation H.222.0 (2006) Information technology—generic coding of moving pictures and associated audio information: systems
ITU-T Recommendation H.262 (2012) Generic coding of moving pictures and associated audio information: video
ITU-T Recommendation H.264 (2012) Advanced video coding for generic audiovisual services
ITU-T Recommendation E.800 (1994) Definitions of terms related to quality of service
ITU-T Recommendation P.800 (1996a) Methods for subjective determination of transmission quality
ITU-T Recommendation P.800 (1996b) Methods for subjective determination of transmission quality
ITU-T Recommendation E.800 (2008) Definitions of terms related to quality of service
ITU-T Recommendation P.810 (1996) Modulated noise reference unit (MNRU)
ITU-T Recommendation P.862: Perceptual evaluation of speech quality (PESQ): an objective method for end-to-end speech quality assessment of narrow-band telephone networks and speech codecs

ITU-T Recommendation P.863 (2011) Perceptual objective listening quality assessment
ITU-T Recommendation P.910 (1999) Subjective video quality assessment methods for multimedia applications
ITU-T Recommendation P.911 (1998) Subjective audiovisual quality assessment methods for multimedia applications
ITU-T Recommendation P.920 (2000) Interactive test methods for audiovisual communications
ITU-T SG12 COM 12–20 (Bellcore, USA) (1993) Experimental combined audio/video subjective test method
ITU-T SG12 D.038. (NTIA/ITS, USA) (1998) Results of an audio-visual desktop teleconferencing subjective experiment
Jones BL, McManus PR (1986) Graphic scaling of qualitative terms. J SMPTE 95:1166–1171
Jones C, Atkinson DJ (1998) Development of opinion-based audio-visual quality models for desktop video-teleconferencing. In: 6th IEEE international workshop on quality of service
Jumisko-Pyykkö S (2011) User-centered quality of experience and its evaluation methods for mobile television. Doctoral dissertation, Tampere University of Technology
Jumisko-Pyykkö S, Hakkinen J, Nyman G (2007) Experienced quality factors—qualitative evaluation approach to audiovisual quality. In: Proceedings of SPIE/IS&T human vision and electronic, imaging (HVEI)
Jumisko-Pyykkö S, Hannuksela MM (2008) Does context matter in quality evaluation of mobile television? In: Proceedings of international conference on human–computer interaction with mobile devices and services (MobileHCI)
Kelly G (1955) The psychology of personal constructs. Norton, New York
Khan A, Sun L, Ifeachor E (2009) Content clustering based video quality prediction model for MPEG4 video streaming over wireless networks. In: Proceedings of IEEE international conference on communications (ICC)
Kohlrausch A, van de Par S (2005) Audio–visual interaction in the context of multi-media applications. In: Blauert J (ed) Communication acoustics, pp 109–138
Korhonen J (2006) New methods for robust audio streaming in a wireless environment. Doctoral dissertation
Lawless HT, Heymann H (1999) Sensory evaluation of food: principles and practices. Chapman & Hall, New York
Le Callet P, Möller S, Perkis A (2012) Qualinet white paper on definitions of quality of experience, version 1.1. In: European network on quality of experience in multimedia systems and services (COST Action IC 1003)
Le Callet P, Péchard S, Tourancheau S, Ninassi A, Barba D (2007) Towards the next generation of video and image quality metrics: impact of display resolution, content and visual attention in subjective assessment. In: Proceedings of of international workshop on image media quality and its applications (IMQA)
Liang Y, Apostolopoulos JG, Girod B (2003) Analysis of packet loss for compressed video: does burst-length matter? In: Proceedings of international conference on acoustics, speech, and signal processing (ICASSP)
Lin TL, Cosman PC, Reibman AR (2008) Perceptual impact of bursty versus isolated packet losses in H.264 compressed video. In: Proceedings of international conference on image processing (ICIP)
List P, Argyropoulos S, Raake A, Feiten B, Garcia MN (2012) Method for estimating the type of the group of picture structure of a plurality of video frames in a video stream. In: WO/2012/013655
List P, Joch A, Lainema J, Bjontegaard G, Karczewicz M (2003) Adaptive deblocking filter. IEEE Trans Circuits Syst Video Technol 13(7):614–619
Liu CM, Hsu HW, Lee WC (2008) Compression artifacts in perceptual audio coding. In: IEEE transactions on audio, speech, and language processing (ASLP)
Liu Y, Kurceren R, Budhia U (2006) Video classification for video quality prediction. J Zhejiang Univ Sci A
Martens JB (2002) Multidimensional modeling of image quality. Proc IEEE 90(1):133–153

Mattila VV (2001) Perceptual analysis of speech quality in mobile communications. Doctoral dissertation
Meltzer S, Moser G (2006) MPEG4 HE-AACv2: audio coding for today's digital media world. In: EBU technical review
Möller S (2000) Assessment and prediction of speech quality in telecommunications. Kluwer Academic Publishers, Boston
Möller A, Raake A, Kitawaki N, Takahashi A, Wältermann M (2006) Impairment factor framework for wide-band speech codecs. In: IEEE transactions on audio, speech, and language processing (ASLP)
Neisser U (1976) Cognition and reality. Principles and implications of cognitive psychology. W.H. Freeman and Company, San Francisco
Offir H (2006) Packet loss concealment for audio streaming. Master thesis
Osgood CE, Suci GJ, Tannenbaum PH (1957) The measurement of meaning. University of Illinois Press
Painter T, Spanias A (2000) Perceptual coding of digital audio. Proc IEE 88(4):451–515
Pan D (1995) A tutorial on MPEG/Audio compression. In: IEEE multimedia
Péchard S, Barba D, Le Callet P (2007) Video quality model based on a spatio-temporal features extraction for H.264-coded HDTV sequences. In: Proceedings of picture coding symposium (PCS)
Perkins C, Hodson O, Hardman V (1998) A survey of packet loss recovery techniques for streaming audio. In: IEEE network
Pervin LA, John OP (2001) Handbook of personality theory and research. The Guilford Press, New York
Pettersson M, Rossholm A (2010) Technique for video quality estimation. WO 2011/082719
Piaget J (1924) Le jugement et le raisonnement chez l'enfant (engl.: Judgement and reasoning in the child. Delachaux & Niestlé S.A., Neuchâtel, Paris
Piaget J (1937) La construction du réel chez l'enfant. (engl.: The child's construction of reality). Delachaux & Niestlé S.A., Neuchâtel, Paris
Pinson MH, Ingram W, Webster A (2011) Audiovisual quality components. IEEE Signal Process 28:60–67
Pinson MP, Wolf S (2003) Comparing subjective video quality testing methodologies. In: Procedings of SPIE visual communications and image processing (VCIP)
Pinson MH, Wolf S (2004) A new standardized method for objectively measuring video quality. IEEE Trans Broadcast 50:312–322
Pinson M, Wolf S (2008) Techniques for evaluating objective video quality models using overlapping subjective data sets. In: NTIA TR-09-457
Raake A (2006) Short- and long-term packet loss behaviour: towards speech quality prediction for arbitrary loss distributions. IEEE Trans Audio Speech Lang Process (ASLP) 14(6)
Raake A (2012) Measuring others' media experience. In: 19th international packet video workshop (PV), Keynote Speaker
Raake A (2006) Speech quality of VoIP—assessment and prediction. Wiley, Chichester
Raake A, Garcia MN, Argyropoulos S, List P, Feiten B (2012) Method and apparatus for assessing the quality of a video signal during encoding and transmission of the video signal. In: WO 2012/076202
Raake A, Garcia MN, Moeller S, Berger J, Kling F, List P, Johann J, Heidemann C (2008) T-V-MODEL: parameter-based prediction of IPTV quality. In: Proceedings of international conference on acoustics, speech, and signal processing (ICASSP)
Raake A, Gustafsson J, Argyropoulos S, Garcia MN, Lindegren D, Heikkilä G, Pettersson M, List P, Feiten B (2011) IP-based mobile and fixed network audiovisual media services. IEEE Signal Process 28
Report EBU (2003) SAMVIQ-subjective assessment methodology for video quality. Report by the EBU Project Group B/VIM, European Broadcasting Union

Richardson IE (2002) H.264 and MPEG-4 video compression: video coding for next-generation multimedia. Wiley, West Sussex

Ries M (2008) Video quality estimation for mobile video streaming. Doctoral Dissertation

Ries M, Crespi C, Nemethova O, Rupp M (2007) Content-based video quality estimation for H.264/AVC video streaming. In: Proceedings of wireless communications and networking conference

Ries M, Puglia R, Tebaldi T, Nemethova O, Rupp M (2005) Audiovisual quality estimation for mobile video services. In: Proceedings of IEEE 2nd symposium on wireless communication systems

Rimac-Drlje S, Vranjes M, Zagar D (2009) Influence of temporal pooling method on the objective video quality evaluation. In: Proceedings of IEEE international symposium on broadband multimedia systems and broadcasting (BMSB)

Rix AW, Beerends JG, Kim DS (2006) Objective assessment of speech and audio quality—technology and applications. IEEE Trans Audio Speech Lang Process (ASLP) 14(6)

Ronsenbluth JH (1998) ITU-T delayed contribution D.064: testing the quality of connections having time varying impairments. Source: AT&T

Rouse DM, Pepion R, Le Callet P, Hemami SS (2010) Tradeoffs in subjective testing methods for image and video quality assessment. In: Proceedings of SPIE/IS&T human vision and electronic, imaging (HVEI)

Rumsey F, Zielinski S, Kassier R, Bech S (2006) On the relative importance of spatial and timbral fidelities in judgments of degraded multichannel audio quality. J Acoust Soc Am 118:968–976

Sanneck H, Carle G (2000) A framework model for packet loss metrics based on loss runlengths. In: Proceedings of SPIE/ACMSIGMM multimedia computer, Networking Conference

Schobben D, van de Par S (2004) The effect of room acoustics on MP-3 audio quality evaluation. In: Proceedings of 117th audio engineering society (AES) convention

Simpson W (2008) Video over IP: IPTV, Internet Video, H.264, P2P, Web TV, and streaming: a Complete guide to understanding the technology. Focal Press Media Technology Professional, Oxford

Spiegel M (1998) Theory and problems of statistics. McGraw Hill, New York

Staelens N, Moens S, Van den Broeck W, Marin I, Vermeulen B, Lambert P, Van de Walle R, Demeester P (2010) Assessing quality of experience of IPTV and video on demand services in real-life environments. IEEE Trans Broadcast 56:458–466

Stevens SS (1946) On the theory of scales of measurement. Science 103:677–680

Strohmeier D, Jumisko-Pyykkö S, Eulenberg K (2011) Open profiling of quality: probing the method in context of use. In: Proceedings of international workshop on quality of multimedia experience (QoMEX)

Strohmeier D, Jumisko-Pyykkö S, Reiter U (2010) Profiling experienced quality factors of audio-visual 3D perception. In: Proceedings of international workshop on quality of multimedia experience (QoMEX)

Takahashi A, Hands D, Barriac V (2008) Standardization activities in the ITU for a QoE assessment of IPTV. IEEE Commun Mag

Tao S, Apostolopoulos J, Guérin R (2008) Real-time monitoring of video quality in IP networks. IEEE Trans Netw

Tektronix. A guide to MPEG fundamentals and protocol analysis. www.tektronix.com/video_audio

Tektronix. An analysis of MPEG encoding techniques on picture quality

Telchemy Incorporated (2008) IP video encoding explained. In: Telchemy application note

Teunissen K, Westerink J (1996) A multidimensional evaluation of the perceptual quality of television sets. J SMPTE

Tourancheau S, Le Callet P, Barba D (2007) Impact of the resolution on the difference of perceptual video quality between CRT and LCD. In: Proceedings of international conference on image processing (ICIP)

Ute Jekosch (2005) Voice and speech quality perception—assessment and evaluation. Springer, Berlin

Verscheure O, Frossard P, Hamdi M (1999) User-oriented QoS analysis in MPEG-2 video delivery. Real-time Imaging
Wältermann M (2012) Dimension-based quality modeling of transmitted speech. Doctoral dissertation, Technische Universität Berlin
Wältermann M, Raake A, Möller S (2010) Quality dimensions of narrowband and wideband speech transmission. Acta Acust United Acous 96:1090–1103
Wältermann M, Raake A, Moeller S (2006) Perceptual dimensions of wideband-transmitted speech. In: 2nd ISCA/DEGA workshop on perceptual quality of systems
Wältermann M, Scholz K, Raake A, Heute U, Moeller S (2006) Underlying quality dimensions of modern telephone connections. In: Interspeech 2006
Wang Z, Lu L, Bovik A (2004) Video quality assessment based on structural distortion measurement. Signal Process Image Commun 19:121–132
Weiss B, Möller S, Raake A, Berger J, Ullmann R (2009) Modeling call quality for time-varying transmission characteristics using simulated conversational structures. Acta Acust United Acust
Wickens CD (1992) Engineering psychology and human performance. Harper Collins, New York
Wiegand T, Sullivan GJ, Bjontegaard G, Luthra A (2003) Overview of the H.264/AVC video coding standard. IEEE Trans Circuits Syst Video Technol (CSVT) 13
Winkler S (1999) A perceptual distortion metric for digital color video. In: Proceedings of SPIE/IS&T human vision and electronic, imaging (HVEI)
Winkler S (2005) Digital video quality. Wiley, Chichester
Winkler S, Faller C (2006) Perceived audiovisual quality of low-bitrate multimedia content. IEEE Trans Multimedia 8(5):973–980
Winkler S, Mohandas P (2008) The evolution of video quality measurement: from PSNR to Hybrid metrics. IEEE Trans Broadcast
Witten IH, Frank E (2005) Data mining. Practical machine learning tools and technique. Morgan Kaufmann Publishers, San Francisco
Wolff T, Ho HH, Foley JM, Mitra SK (2006) H.264 coding artifacts and their relation to perceived annoyance. In: Proceedings of EUSIPCO
Wu HR, Rao KR (2006) Digital video image quality and perceptual coding. Taylor & Francis, Boca Raton
Wu H, Claypool M, Kinicki R (2006) Guidelines for selecting practical MPEG group of pictures. In Proceedings of IASTED international conference on internet and multimedia systems and applications (EuroIMSA)
Yamada T, Yachida S, Senda Y, Serizawa M (2010) Accurate video-quality estimation without video decoding. In: Proceedings of international conference on acoustics, speech, and signal processing (ICASSP)
Yamagishi K, Hayashi T (2009) Non-intrusive packet-layer model for monitoring video quality of IPTV services. IEICE Trans Fundam 12:3297–3306
Yamagishi K, Hayashi T (2005) Analysis of psychological factors for quality assessment of interactive multimodal service. In: Proceedings of SPIE/IS&T human vision and electronic imaging (HVEI), vol 5666–15, pp 130–138
Yamagishi K, Hayashi T (2008) Parametric packet-layer model for monitoring video quality of IPTV services. In: Proceedings of IEEE international conference on communications (ICC)
You J, Reiter U, Hannuksela MH, Gabbouj M, Perkis A (2010) Perceptual-based quality assessment for audio-visual services: a survey. J Signal Process: Image Commun
You F, Zhang W, Xiao J (2009) Packet loss pattern and parametric video quality model for IPTV. In: Proceedings of international conference on computer and information Sci (ICIS)
Yuen M, Wu HR (1998) A survey of hybrid MC/DPCM/DCT video coding distortions. Signal Process, EURASIP J 70
Zacharov N, Koivuniemi K (2001) Audio descriptive analysis and mapping of spatial sound displays. In: Proceedings of the 2001 international conference on auditory display
Zieliński S, Rumsey F, Bech S (2008) On some biases encountered in modern audio quality listening tests-a review. In: Proceedings of 124th audio engineering society (AES) convention
Zou WY, Corriveau PJ (1996) Methods for evaluation of digital television picture quality. J SMPTE

MIX
Papier aus verantwortungsvollen Quellen
Paper from responsible sources
FSC® C105338
FSC
www.fsc.org

Printed by Libri Plureos GmbH
in Hamburg, Germany